Notice

Quality Assurance Statement

Table of Contents

List of Tables

List of Figures

1.0   Introduction

The ability of our nation's transportation system to provide for and maintain the efficient movement of freight is important to the continuing economic health of the United States. U.S. domestic freight tonnage is anticipated to approximately double – and international freight tonnage expected to nearly triple – by 2035. This has led to a growing need to find new ways to address air quality concerns and greenhouse gas emissions associated with freight movements.

Diesel exhaust from freight vehicles is a primary source of $PM_{2.5}$, air toxic contaminants, and NO_x emissions (one precursor to ozone), all of which have potential health implications. Freight emissions comprise close to one-third of U.S. transportation greenhouse gas emissions, and have grown by more than 50 percent since 1990. As a result, there is a steadily increasing number of challenges faced by both freight and air quality planners as they attempt to simultaneously meet the growing demand for freight while improving environmental outcomes. They must meet the requirements of new and varied initiatives being put into place across the nation as states and regions grapple with air quality issues and emissions budgets, understand how to integrate emerging equipment and infrastructure technologies, look for ways to make the system more efficient, and identify new funding sources for these activities.

A wide range of strategies is available to mitigate these freight and air quality challenges, ranging from technological strategies such as engine retrofits and alternative fuels, to operations strategies such as congestion mitigation and idling reduction. Not only must planners identify viable mitigation strategies, they also must navigate the myriad transportation and environmental funding programs to identify ones that could be applicable to their project.

The Federal Highway Administration's (FHWA) Office of Freight Management and Operations in cooperation with the Office of Natural and Human Environment developed this handbook as a resource for states, metropolitan planning organizations (MPO), FHWA and other public- and private-sector organizations to use in developing solutions to these challenges. This handbook provides the background needed to understand how freight contributes to air quality issues, describes strategies to mitigate those freight-related pollutant emissions and improve air quality, and identifies funding and financing tools available for freight-related air quality projects (e.g., freight projects designed to reduce the emissions of air pollutants). It is designed to be used by all involved in the identification, financing, and delivery of freight-related air quality projects, whether approaching from a freight or an air quality perspective.

Four sections follow this introduction:

- **Section 2.0: Background on Freight-Related Air Quality** provides an introduction to the nature of freight-related pollutants and associated air quality impacts, the sources of emissions by freight mode, and the conditions, laws, and regulations that govern air quality and can impact freight operations and investment decisions.

- **Section 3.0: Strategies for Freight Transportation-Related Emission Reduction/Air Quality Improvement Projects** describes the strategies available to reduce emissions from freight movements, from technology applications to operational, policy, and regulatory initiatives.

- **Section 4.0: Funding and Financing Tools for Freight Air Quality Improvements** describes the major funding and financing options available for freight-related air quality improvements, focusing particularly on Federal options.

- **Section 5.0: Case Studies** describes notable freight projects and programs that are meant to improve air quality and reduce freight-related emissions, to provide real-world examples of how strategies and programs are applied in practice.

2.0 Background on Freight-Related Air Quality

Understanding the air quality issues and regulatory environment related to the freight sector is an important component of planning for transportation projects and specifically freight-related air quality projects (freight projects designed to reduce air pollutant emissions). This section provides background on the nature of freight-related pollutants and associated air quality impacts, the sources of freight emissions by mode, and the conditions, laws, and regulations that govern air quality and impact freight operations and investment decisions. It will provide transportation and freight practitioners less versed in air quality with basic information about air quality rules, regulations, and impacts, allowing them to better communicate with resource agency staff and environmental professionals. For those with broader air quality planning experience, it provides background information on the types of air quality impacts that are unique to freight movements and operations.

2.1 EMISSIONS AND AIR QUALITY IMPACTS

This section provides an overview about various types of air pollutants associated with the freight transportation sector, the health and environmental impacts of each, their impacts on regional and local air pollution and global issues (climate change), and some of the tools and methods used to calculate emissions. Specific types of air pollutants covered include:

- **Criteria Pollutants** – Six key pollutants for which the U.S. Environmental Protection Agency (EPA) establishes national ambient air quality standards (carbon monoxide, nitrogen dioxides, ozone, particulate matter, sulfur dioxide, and lead). Nitrogen dioxides, particulate matter, and ozone are the pollutants of greatest concern for the U.S. freight transportation sector;

- **Mobile-Source Air Toxics (MSAT)** – Compounds with significant contributions from mobile sources for which there are no established standards; and

- **Greenhouse Gases (GHG)** – Gaseous compounds that trap heat in the earth's atmosphere and contribute to global climate change. Carbon dioxide (CO_2) is the primary greenhouse gas emitted by freight vehicles.

2.1.1 CRITERIA POLLUTANTS

Criteria pollutants are a group of air pollutants for which the U.S. Environmental Protection Agency (EPA) sets National Ambient Air Quality Standards (NAAQS), following the requirements of the Clean Air Act.1 The EPA sets national guidelines ("criteria") for permissible levels of criteria pollutants, based on scientific knowledge of their human health and/or environmental impacts. The six criteria pollutants regulated by the EPA are:

- **Carbon monoxide (CO)** is a gas that is formed when the carbon in a fuel source is not burned completely. Nationally, motor vehicle exhaust accounted for about 54 percent of CO emissions in 2005.[2] In large cities, the proportion can be much higher due to the concentration of population and vehicle-miles traveled. Other sources of CO emissions include industrial processes like metal processing, residential wood burning, gas stoves, and cigarette smoke. Even low-level exposure to CO can have harmful cardiovascular effects, particularly for people who suffer from heart disease. Exposure to high levels of carbon monoxide can have effects on the central nervous system such as vision problems, reduced ability to learn or perform complex tasks, and reduced dexterity. Goods movement activities are not a significant source of CO pollution, as diesel engines are not major emitters of CO.

- Nitrogen dioxides (NO_2) are a family of reactive gaseous compounds that contribute to air pollution in both urban and rural areas. They are produced during the combustion of fuels at high temperatures. Transportation produced 59 percent of NO_x emissions in 2005; electricity generation is the next largest contributor. Freight transport (heavy-duty trucks, marine, rail, and air cargo) accounted for approximately 57 percent of transportation emissions, with heavy duty trucks and buses accounting for 26 percent, marine vessels for 22 percent, locomotives nine percent, and freight aircraft less than one percent.[3] While EPA's National Ambient Air Quality Standard covers this entire group of NO_x, NO_2 is the component of greatest interest and used as the indicator for the larger group of nitrogen oxides for the purposes of regulation. NO_x also is a precursor[4] to other pollutants;

---

[1] 40 CFR Part 50.

[2] Environmental Protection Agency, 2005 National Emissions Inventory.

[3] Ibid. Freight sources include heavy duty trucks and buses, marine vessels, railroads, heavy duty gasoline vehicles, and aircraft emissions attributable to cargo operations.

[4] A precursor is a primary pollutant that turns into a criteria pollutant, either through chemical reaction or decay.

for example, it reacts with volatile organic compounds (VOC) in the presence of sunlight to form ozone. It also reacts with sulfur dioxide to form acid rain, which can raise the acidity of water bodies and make them unsuitable for many fish.

- **Ground-level ozone (O_3)** is not actually emitted, but is formed from a chemical reaction between NO_x and VOCs. Sunlight breaks down the precursor chemicals (NO_x and VOC) in a process called photolysis, after which oxygen atoms combine to form ozone. As a result, ozone concentrations tend to be highest in the summer when there are more sunny days. Sources of the pollutants that create ozone include vehicle exhaust, industrial processes, gasoline vapors, chemical solvents, and even some elements of natural vegetation. Ground-level ozone is the primary component of smog. Ozone formation can be exacerbated by congested daytime operations at major freight facilities, such as ports, since these facilities are typically busiest during daylight hours. Wind currents can carry ozone and the pollutants that form it for many miles, so even areas with low freight volumes can be affected. Ozone has been linked to respiratory problems, sunburn-like skin irritation, aggravation of asthma, and increased susceptibility to pneumonia and bronchitis, as well as reduced crop yields and damage to vegetation. Reliable estimates of the proportion of ozone that is attributable to freight do not exist, but diesel engines are a significant source of NO_x, which is a precursor to ozone.

- **Particulate matter (PM)** is composed of small particles and liquid droplets of a variety of chemicals and other agents, such as dust particles, organic chemicals, acids, and metals. Some PM is directly emitted from the tailpipe as a by-product of engine combustion. Secondary PM, on the other hand, is formed by reactions in the exhaust plume outside the vehicle when fine PM molecules attach to other molecules (nucleation) or to each other (homogeneous nucleation). In addition, road dust is a major component of PM. Particulate matter aggravates asthma symptoms and has been linked to cancer and heart disease, chronic bronchitis, irregular heartbeat, nonfatal heart attacks, and premature death in individuals with heart or lung disease. For regulatory purposes, particulate matter is grouped into two categories based on the size of the particles:

  - **PM_{10}** is made up of particles less than 10 microns in diameter (about one-seventh the width of a human hair). Nationally, road dust (including tire particles and brake dust) is the largest source of PM_{10} emissions. Road dust accounted for almost 11 million tons of PM_{10} emissions in the United States in 2005, over half of the total.[5] The transportation sector (including road dust) is responsible for about 54 percent of PM_{10} emissions. Freight movements produce 51 percent of transportation emissions, with marine vessels accounting for 29 percent of transportation emissions, heavy duty trucks and buses for 17 percent, and locomotives for five percent. Major non-transportation sources include fires, agriculture, and electricity generation.

- $PM_{2.5}$ is composed of particles less than 2.5 microns in diameter; motor vehicle exhaust is a major source of this type of pollution. There is evidence that $PM_{2.5}$ is more hazardous to human health than the larger PM_{10} particles, largely because they can travel in the air for very long distances, tend to remain in the lungs when inhaled, and can even enter the bloodstream. Fine PM is often created through secondary formation when diesel exhaust particles react with other compounds in the atmosphere such as NO_x. Together, on-road vehicles and non-road equipment (including rail and marine sources) contributed about 10 percent of total $PM_{2.5}$ emissions in 2005, or about 550,000 tons. Road dust made up another 1.2 million tons (21 percent).[6]

• **Sulfur dioxide (SO_2)** is a gas formed when fuel sources containing sulfur (such as coal and oil) are burned, and when crude oil is converted to gasoline. These gases dissolve into water easily; in fact, SO_2 combines with water vapor in the atmosphere to help form acid rain. SO_2 also is associated with breathing difficulties and respiratory illness, contributes to haze in the air, and damages crops and other plant life. Electricity generation (mostly by coal-fired power plants) is the largest generator of SO_2 emissions nationwide, accounting for over 67 percent of total emissions in 2005. All transportation sources combined accounted for 11 percent of the total.[7]

• **Lead (Pb)** is a naturally occurring metal found in the environment and in many manufactured products. It can be inhaled from the air or ingested from contaminated drinking water/food, or from lead-based paint found in older buildings. Once ingested, it enters the bloodstream and can accumulate in the bones. Lead has numerous negative effects on the respiratory, cardiovascular, reproductive, and immune systems. Children are especially sensitive to lead exposure. In the environment, lead causes neurological effects in vertebrates,

---

[5] Environmental Protection Agency, *2005 National Emissions Inventory*. Since the 2005 NEI does not identify road dust separately from other fugitive dust, it was estimated using percentages from 2002.

[6] Ibid.

[7] Environmental Protection Agency, *2005 National Emissions Inventory*.

decreased growth and reproduction in plants and animals, and a general loss of biodiversity. Historically, the transportation sector was a major source of lead pollution, but the elimination of lead in gasoline led to a 95 percent drop in transport-related lead emissions between 1980 to 1999. As a result of this regulatory action, on-road vehicles are no longer a significant source of lead pollution. However, non-road equipment (including locomotives, ships, and planes) are still a significant source of lead pollution. In 2002, nearly 28 percent (464 tons) of lead emissions came from these sources.[8] Fuel containing lead is still sold in small amounts for specific applications such as race cars, farm equipment, and aircraft.

2.1.2 MOBILE-SOURCE AIR TOXICS

MSATs are pollutants emitted from highway vehicles and non-road equipment. MSATs may have serious health effects, but unlike criteria pollutants they are not regulated by NAAQS. The seven MSATs of particular concern are acrolein; benzene; 1,3-butadiene; formaldehyde; diesel particulate matter and diesel exhaust organic gases; naphthalene; and polycyclic organic matter.[9] Some of these chemicals (such as benzene) are present in gasoline and diesel fuel and are emitted through evaporation or when fuel passes through an engine without being burned. Others (such as formaldehyde and diesel particulate matter) are not present in the fuel itself; rather, they are byproducts of incomplete combustion. These compounds have a variety of potential human health effects. Benzene, for instance, is a known carcinogen, while formaldehyde and diesel particulate matter are probable carcinogens. An EPA study concluded that long-term inhalation of diesel exhaust probably poses a lung cancer risk and can cause other respiratory problems.[10]

2.1.3 GREENHOUSE GASES

Greenhouse gases (GHG) are gaseous compounds that trap heat in the earth's atmosphere. They can be naturally occurring or man-made. There are several greenhouse gases that are the result of human activity, but carbon dioxide (CO_2) is the primary concern from a freight perspective, since it is formed through the burning of fossil fuels such as oil, natural gas, and

---

[8] U.S. Environmental Protection Agency, 'National Summary of Lead Emissions,' Retrieved December 16, 2008 from http://www.epa.gov/air/emissions/pb.htm.

[9] U.S. Environmental Protection Agency, "Control of Hazardous Air Pollutants from Mobile Sources; Final Rule," *Federal Register* Volume 72, Number 37, Monday February 26, 2007: pages 8427-8570.

[10] United States Environmental Protection Agency, *Health Assessment Document for Diesel Engine Exhaust*, May 2002.

coal. In 2007, the transportation sector produced about 32 percent of the nation's CO_2 emissions, 60 percent of which was attributable to passenger vehicles. Much of the remainder came from freight sources.[11] Unlike criteria pollutants or MSATs, greenhouse gases are global in nature and can remain in the atmosphere for very long periods of time (50 to 200 years in the case of CO_2).

GHGs are the cause of the "greenhouse effect," which refers to the rise in earth's temperature that results from atmospheric gases trapping the sun's heat. As such, GHGs (including CO_2) are a primary contributor to global warming since they enhance the heat-trapping properties of the atmosphere. Since the Industrial Revolution, the concentration of carbon dioxide in earth's atmosphere has risen by about 30 percent, largely due to human activities. During the last 100 years, the global average surface temperature has risen by 1.33 degrees Fahrenheit; most of that increase (1.17 degrees Fahrenheit) occurred in the last 50 years.[12] Increased concentrations of GHGs will likely accelerate this trend.

Climate change can be linked to other environmental phenomena like sea-level rise, increased precipitation, and increased hurricane intensity. The recent warming of earth's atmosphere has been linked to melting sea ice (which affects currents and ecosystems) as well as melting glaciers and ice sheets on land (which raises the sea level). About one-third of the CO_2 generated by the burning of fossil fuels is absorbed into the ocean, where it raises the acidity of surface water, which in turn can have potential negative effects on marine life.[13]

Besides the risks to life and property, climate change can adversely affect freight movements and transportation in general.[14] For example, it has been projected that an 18-foot storm surge (such as that produced by a hurricane) would inundate 41 percent of rail miles operated, 64 percent of Interstate miles, and 57 percent of arterial highway miles along the U.S. Gulf Coast.[15]

---

[11] EPA, *Inventory of U.S. Greenhouse Gas Emissions and Sinks: 1990 to 2007*, April 2009.

[12] Intergovernmental Panel on Climate Change Working Group I, *Assessment Report 4*, November 2007.

[13] Grimond, J. "Troubled waters." *The Economist*, December 30, 2008.

[14] Global warming may also enhance freight movement. There is a chance that melting Arctic sea ice will open up new shipping lanes through the Arctic Ocean, which would cut 2,000 miles off of a trip from Rotterdam to Seattle as compared to transiting through the Panama Canal.

[15] Climate Change Science Project, *Potential Impacts of Climate Variability and Change on Transportation Systems and Infrastructure – Gulf Coast Study*, March 12, 2008.

2.1.4 EMISSIONS MODELING

There are a number of quantitative tools that air quality practitioners and transportation planners can use to estimate freight vehicle emissions and model their impacts. The EPA's MOVES2010[16] model estimates emissions of VOC, NO_x, PM (PM_{10} and $PM_{2.5}$), CO, MSATs (benzene; 1,3-butadiene; formaldehyde; acetaldehyde; acrolein; naphthalene; ethanol; and MTBE), and greenhouse gases for cars, trucks, buses, and motorcycles outside California. In California, emissions analyses are conducted using the EMFAC2007 model. This model was developed by the California Air Resources Board and approved by EPA.

Estimating emissions from freight movements also should consider non-road sources, like locomotives, ships, and aircraft (although non-road sources are not considered in the transportation conformity process described later in Section 2.3.1). This is especially important in communities with ports or significant rail traffic. EPA publishes emissions rates and methodologies for different types of rail movements (line-haul, short-line, switch, etc.) as well as aircraft. EPA also publishes guidance on estimating marine vessel emissions. EPA's NONROAD model can be used to estimate emissions from non-road sources other than ships, trains, and aircraft. This would include cargo handling equipment like port gantry cranes, forklifts, and container handlers.

Other EPA modeling tools include:

- The Freight Logistics Environmental and Energy Tracking (FLEET) model is a spreadsheet-based modeling tool to help truck fleet owners optimize fuel economy and reduce emissions. The model helps fleet managers track fuel economy and estimate how much CO, NO_x, and PM emissions they can prevent through various measures.

- The National Mobile Inventory Model (NMIM) is a desktop computer application that helps planners and air quality analysts develop estimates of current and future emissions inventories from mobile sources, including on- and off-road freight vehicles. The on-road emissions calculations of NMIM is based on MOBILE6.2. NMIM can be used to calculate national, state, or county-level inventories.

- The DrayFLEET model is a spreadsheet-based model that calculates emissions from container drayage activities at ports. Drayage trips are truck trips to move containers within port complexes and to and from intermodal transfer facilities and depots. DrayFLEET allows

---

[16] Further information regarding MOVES2010 can be found on EPA's web site at http://www.epa.gov/otaq/models/moves/index.htm.

port planners to understand the impact on emissions of changing management practices, terminal operations, cargo volume, and technology upgrades.

- Fuels models can be used to estimate the emissions impacts of changes in fuel properties and composition. EPA has produced technical reports on the effects of different diesel fuel formulations on emissions.

2.2 FREIGHT SOURCES OF AIR POLLUTANTS

Most freight emissions are from diesel engines, and diesel exhaust is a major source of PM, NO_x, and SO_x pollutants. This section will focus on the characteristics that differentiate among the modes (fuels, mode of operation, etc.). The different modes include trucks, marine vessels and ports, rail vehicles, and air cargo.

2.2.1 TRUCK

Trucks remain the most dominant mode for freight movements, by weight, value, and ton-miles. In 2007, trucks carried about 61 percent of total freight tonnage in the United States, and more than 65 percent of total freight value.[17] The EPA has introduced stringent new caps on emissions of PM, NO_x, and other pollutants for model year 2007 and later trucks. At the same time, it mandated the use of ultra-low sulfur diesel (ULSD) in heavy-duty trucks produced since the 2007 model year, which enables the use of more advanced pollution control technology in diesel engines (higher sulfur fuels can "poison" the catalyst used in NO_x and PM aftertreatment technologies). To support that rule, refiners began producing ULSD in mid-2006, and its use was required for the on-road fleet in 2007. However, there are still millions of older trucks in the nation's fleet that lack these new aftertreatment technologies. In any case, many of these trucks were built before the most recent emissions standards for diesel engines came into effect, which limits the effect of the new standards in the short term. Vehicle maintenance also can affect truck emissions; poorly maintained trucks often emit more than those that are kept in good running condition. Unlike locomotives, trucks are not subject to engine rebuild emissions standards.

It should be noted that 10 percent of medium- and heavy-duty truck fuel consumption is from gasoline trucks. Emissions from gasoline-powered freight vehicles are an important contributor to freight-related emissions.

---

[17] Federal Highway Administration, Freight Analysis Framework.

2.2.2   MARINE VESSELS AND PORTS

Marine cargo vessels and port complexes are the second largest source of diesel freight emissions. In addition to cargo ships, ports use cranes, hostlers, and other equipment powered by diesel fuel. Cargo vessels typically burn bunker fuel (also known as residual fuel because it is literally left over from the refining process), a form of diesel fuel with particularly high sulfur content. Bunker fuel is the most common fueling option because of its low cost; considering that a typical cargo ship burns 120 gallons of fuel per mile.[18] They are major contributor to air quality issues in coastal regions, especially those related to sulfur oxides. Researchers have estimated that ships burning this type of fuel are responsible for as many as 60,000 deaths per year worldwide and cost the U.S. economy about $500 million annually.[19] Another study found that as much as 44 percent of the primary sulfates (a very small particulate matter found in diesel exhaust) in the air in California coastal areas comes from ships.[20]

Marine diesels are classified by the EPA into Category I, II, and III engines. Category III engines are the very large engines used on oceangoing cargo vessels such as containerships. These engines are the primary users of bunker fuel. Category I and II engines generally burn cleaner distillate fuel. However, these engines are subject to less stringent emissions regulations than diesel engines designed for on-road use.

In addition to the ships themselves, cargo handling equipment at ports are often powered by diesel engines. These can include container cranes[21] (which offload cargo containers from ships for transfer to trucks or trains), forklifts, terminal tractors, and container handlers, among other things. Although many of these vehicles utilize clean-diesel technologies (and many ports mandate the use of such technology), they still contribute to air quality issues around ports.

---

[18] Barry, K. "Toyota's Solar Car Carrier." Wired Blog Network, September 3, 2008.

[19] Corbett, J., Winebrake, J., Green, E., Kasibhatla, P., Eyring, V., and Lauer, A. "Mortality from Ship Emissions: A Global Assessment." *Environmental Science and Technology*, November 5, 2007.

[20] McDonald, K. "Dirty Smoke from Ships Found to Degrade Air Quality in Coastal Cities." University of California, San Diego News Center, August 18, 2008.

[21] Most modern container cranes use electric power, but there are still many older diesel-powered cranes in use.

2.2.3 RAIL

Rail locomotives are another source of significant diesel exhaust pollution. There currently are 20,000 freight locomotives in use across the country.[22] Rail is often held up as a clean alternative to trucks, and it is true that one train can move an equivalent volume of 250 trucks, and that trains enjoy about a three to one advantage in fuel efficiency (emissions per ton/mile) over trucks. However, emissions standards for locomotives lag behind those for trucks, and many older locomotives that are still in use predate even the most basic regulations (more stringent locomotive emissions requirements are slowly being phased in and are described in Section 2.3.2). Locomotives have 30- to 40-year service lives, so older, more polluting models remain in use longer than trucks typically do, although engines are typically rebuilt every 600,000 to 1,000,000 miles.[23] Like trucks, trains emit significant amounts of NO_x and particulate matter.

In addition, rail freight is growing for a number of reasons. These include escalating fuel costs (which plays to rail's fuel efficiency advantage) and a shortage of truck drivers. The U.S. Department of Transportation estimates that total rail freight tonnage will grow by 73 percent between 2006 and 2035.[24] The EPA estimates that without new controls, locomotives and ships will contribute 27 percent of total mobile source NO_x and 45 percent of mobile source fine diesel particulate matter ($PM_{2.5}$) emissions by 2030.[25]

Freight rail locomotives fall into two groups: 1) line-haul locomotives; and 2) switchyard locomotives. Line-haul locomotives are the more powerful engines that the railroads use to move large freight trains between major hubs. Switchyard locomotives are less powerful and are used to disassemble and reassemble trains by moving cars around at a rail yard. Line-haul operations, involve a greater proportion of operating time at high power levels, while locomotives engaged in switching operations typically spend most of their time at a lower power output, starting and stopping, or at idle. This tends to increase emissions for switchers, since frequent acceleration and deceleration requires more power than cruising at a constant speed. Railroads also tend to "sunset" older locomotives by shifting them from line-haul duty to switchyard functions, meaning that rail yards (sometimes located in dense urban population centers) often

---

[22] Palaniappan, M., Prakash, S., and Bailey, D. *Paying With Our Health: The Real Cost of Freight Transport in California.* The Pacific Institute, November 2006.

[23] Stodolsky, F. (2002). *Railroad and Locomotive Technology Roadmap.* Argonne National Laboratory, Center for Transportation Research, ANL/ESD/02-6.

[24] Federal Highway Administration, Freight Analysis Framework.

[25] http://www.epa.gov/nonroad-diesel/420f04041.htm.

end up with the oldest, most polluting locomotives; however, these loco-motives are subject to updated engine-rebuild emissions standards when they go through major overhauls.

2.2.4 AIR FREIGHT

Air cargo is a very small part of total freight movements in the United States, when measured by weight. This is because moving goods by air is very expensive. Generally, light, higher-value, more time-sensitive com-modities move by air. In fact, despite being less than one percent of total freight volumes in 2006, air cargo movements comprised seven percent of total freight value that year.[26] Air cargo movements are expected to grow faster in volume than other modes (with growth rates of up to four percent annually by some estimates[27]), so they will likely contribute more to air quality problems in the future, especially in large urban areas with major airports. In addition, air movements almost always require a truck trip on either side of the shipment (air-rail moves are possible, but rare), so truck volumes and their associated emissions grow along with air cargo moves. Cargo and baggage handling equipment at airports primarily serves passenger jets, but also is a source of airport-related emissions.

It is hard to isolate air freight emissions because a large proportion of air cargo is carried in the cargo holds of commercial passenger aircraft. The FHWA estimates that just 0.1 percent of NO_x and 0.2 percent of PM_{10} emissions related to freight come from air cargo operations. However, in some cities the proportion is much higher; in Los Angeles, for example, air freight accounts for 0.5 percent of total freight NO_x emissions and 0.3 percent of freight PM_{10} emissions.[28] All jet aircraft (passenger and freight) emit VOCs, NO_x, SO_2, and CO. Aircraft operations that occur below 3,000 feet are considered to have an impact on ground-level air quality. Like locomotives, commercial jets have long service lives (25 to 40 years), so it can take decades before technological improvements or new regulatory standards that reduce emissions show up in the majority of the fleet.

2.2.5 TOTAL EMISSIONS BY MODE AND ECPECTED TRENDS

Trucks accounted for 46 percent of freight-related NO_x emissions in 2005, the largest share of any single mode (Figure 2.1). While trucks' share of total emissions for both pollutants has declined in recent years due to the advent of stricter emissions regulations, they still account for a significant

---

[26] Federal Highway Administration, Freight Analysis Framework.

[27] FHWA and ICF Consulting, *Assessing the Effects of Freight Movement on Air Quality at the National and Regional Level,* April 2005.

[28] Ibid.

amount of freight emissions because most freight in the United States moves by truck. Marine vessels made up the next largest share at 38 percent, followed by railroads (16 percent) and air cargo (less than 1 percent). In contrast, ships comprised 57 percent of PM_{10} emissions in 2005, compared to 34 percent for trucks, 9 percent for rail, and less than 1 percent for air cargo.

Figure 2.1   Total NO_x and PM_{10} Emissions by Mode 2005

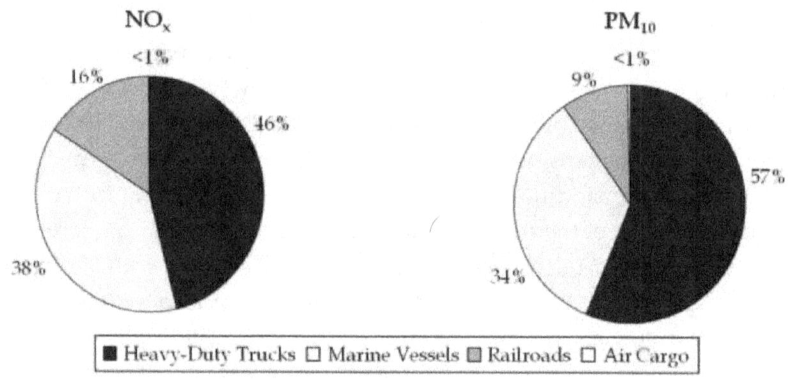

Source: EPA 2005 National Emissions Inventory.

The share of emissions by mode can vary significantly by region. For example, over 23,000 tons of NO_x in the Chicago region in 2002 (19 percent of total freight emissions) came from freight rail, reflecting that region's status as a major North American rail freight hub. Similarly, marine operations accounted for a large proportion of PM_{10} freight emissions in Houston (40 percent) and Los Angeles (37 percent), a result of the major port facilities present in those cities.[29]

With the exception of air freight, most of the modal sources of freight pollution are expected to decline in the future (Figure 2.2).[30] These reductions will largely be the result of stricter EPA regulations governing mobile source emissions. Heavy-duty truck emissions are expected to decline the most (by about 82 percent), with freight rail emissions declining by 43 percent. Marine emissions are expected to decline much more modestly due to less stringent regulations and the fact that most cargo vessels calling on U.S. ports are foreign flagged and, therefore, not subject

---

[29] FHWA and ICF Consulting, *Assessing the Effects of Freight Movement on Air Quality at the National and Regional Level,* April 2005.

[30] This forecast uses a 2002 base year. With the gradual phasing in of stricter regulations beginning in 2007 (particularly for truck and rail), the declines would likely be more modest if a more recent base year were used, but would still be substantial given fleet turnover rates.

to EPA regulations. Air freight PM_{10} emissions are expected to decline slightly, but NO_x emissions related to air cargo will increase by 51 percent by 2020. Overall, PM_{10} emissions from freight sources are expected to decline by 5 percent annually through the year 2020, resulting in a 63 percent total decline; NO_x emissions will decline by 4 percent per year (50 percent overall).

Figure 2.2   Future Freight-Related NO_x and PM_{10} Emissions Change by Mode
2010 and 2020

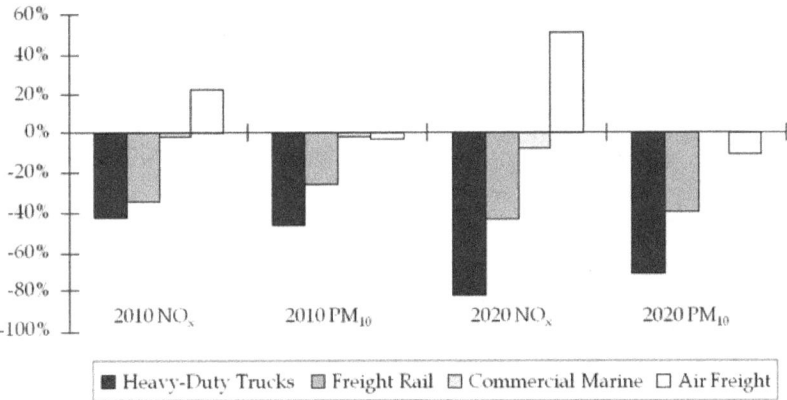

■ Heavy-Duty Trucks   ▨ Freight Rail   ☐ Commercial Marine   ☐ Air Freight

2.3   REGULATORY ENVIRONMENT

This section provides an overview of Federal, state, and local laws and regulations dealing with emissions and air quality, as well as agency roles and responsibilities. Each level of government has a distinct set of responsibilities related to transportation and environmental protection, as shown in Figure 2.3.

- **The Federal government** sets national policy and performs important oversight roles. The EPA, for example, develops national standards for certain pollutants as noted below. In the case of transportation, the Federal government (FHWA and FTA) also provides funding and financing for projects.

- **State and regional agencies** are normally responsible for carrying out Transportation plans, programs, and policies. For air quality policy, the EPA sets national standards, but it is up to state resource agencies to determine how best to comply with them. Some states set their own air quality standards that are more stringent than the Federal standards. Similarly, state DOTs plan and carry out transportation projects at the statewide level.

- **Local agencies** also can set their own standards for air quality that are equal to or more stringent than state standards. Metropolitan Planning Organizations (MPO) determine how Federal transportation funds will be spent in their areas; they also must ensure that transportation plans and programs conform to the purpose of the state implementation plans. Many local governments also have adopted laws to control freight-related emissions.

Figure 2.3   Agency Roles and Responsibilities

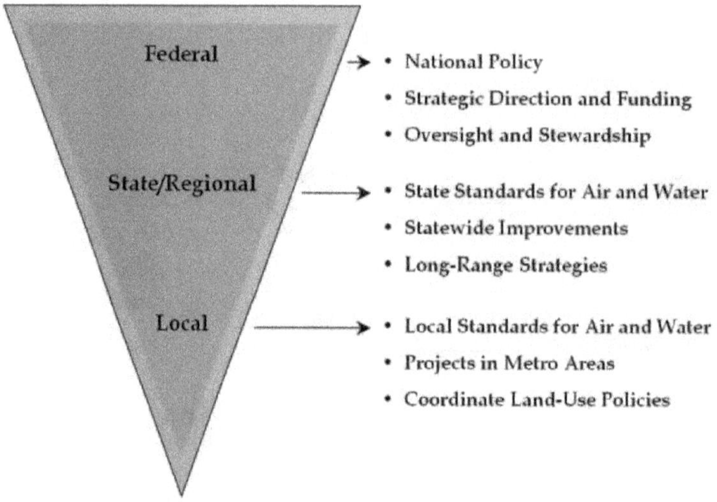

Federal → • National Policy
• Strategic Direction and Funding
• Oversight and Stewardship

State/Regional → • State Standards for Air and Water
• Statewide Improvements
• Long-Range Strategies

Local → • Local Standards for Air and Water
• Projects in Metro Areas
• Coordinate Land-Use Policies

The following sections discuss the Federal, state, and local roles in more detail to provide a complete picture of the regulatory environment affecting freight and air quality.

2.3.1   CLEAN AIR ACT

The Clean Air Act (CAA) was first enacted in 1963 and has been revised many times since then. The Clean Air Act of 1970 represented a major shift in Federal pollution control activities by authorizing comprehensive requirements for the control of stationary, area and mobile source emissions. The most recent revisions (in 1990), often called the CAA Amendments or CAAA, increased the authority and responsibilities of the Federal government.

Air Quality Standards

As discussed in Section 2.1.1, the national ambient air quality standards (NAAQS) set limits for six criteria pollutants: CO, NO_2, O_3, PM, SO_2, and lead. The CAA requires that areas exceeding the limits for one or more of the criteria pollutants (as measured through air quality monitoring) be designated as nonattainment. States that have nonattainment areas are required to develop a State Implementation Plan (SIP), which is a detailed description of the resources and programs a state will use to achieve and maintain NAAQS. The SIP is based on analytical methods approved by the EPA and is developed in consultation with local transportation and resource agencies. A state's SIP can incorporate freight issues; for example, the Texas SIP for the Houston-Galveston region includes a transportation control measure that focuses on NO_x reductions and includes a voluntary emissions program for railroads.

Areas which were previously designated as nonattainment but have consistently met the NAAQS over a 3-year period with no violations on their monitors are redesignated as attainment and called maintenance areas. Maintenance areas must develop maintenance SIPs which show how the area will maintain the NAAQS for two 10-year periods.

The CAA requires the Federal government to conform to air quality goals in the SIP before approving or funding any activity.[31] Conformity is the process used to meet this requirement. Freight activities are either covered under transportation or general conformity, depending on which Federal agency is funding and/or approving the project. Transportation conformity only applies to on-road mobile sources – not freight rail, marine, or aviation – in nonattainment and maintenance areas for transportation-related pollutants (ozone, CO, PM_{10}, $PM_{2.5}$ and NO_2). It specifically applies to metropolitan transportation plans, programs and projects developed, funded, or approved under title 23 U.S.C. or Federal Transit Laws. General conformity applies in nonattainment and maintenance areas for all criteria pollutants. It also applies to all other Federal actions not covered under transportation conformity, such as airports and railroads. Interagency consultation should be used to determine how the conformity requirements are met for a particular freight project.

Mobile Source Air Toxics (MSAT)

MSAT analysis may be relevant in some cases. In February 2007, EPA promulgated new regulations to reduce MSATs by limiting the amount of benzene in gasoline and reducing emissions from passenger vehicles and gas cans. The FHWA has published interim guidance for project sponsors conducting MSAT analysis, which is encouraged but not required for environmental documentation on Federally funded transportation projects.[32] FHWA uses a three-tiered approach with projects divided into groups:

---

[31] Transportation conformity is covered under 40 CFR Parts 51 and 93, Subpart A, while general conformity is covered under 40 CFR Parts 6, 51, and 93, Subpart B.

[32] FHWA, 2009: "INFORMATION: Interim Guidance Update on Mobile Source Air Toxic Analysis in NEPA Documents." Memorandum from April Marchese, Director, Office of Natural and Human Environment, September 30, 2009. http://www.fhwa.dot.gov/environment/airtoxic/100109guidmem.htm.

- **Projects not requiring analysis** are those with no potential for meaningful MSAT effects. These projects include those that are categorically excluded under 23 CFR 771.117(c); are exempt under the transportation conformity rules per 40 CFR 93.126; or others that have no meaningful impact on traffic volumes or vehicle mix.

- **Projects requiring qualitative analysis** are those projects with low potential MSAT effects. These projects include those that serve to improve operations of highway, transit, or freight without adding substantial new capacity or without creating a facility that is likely to meaningfully increase emissions.

- **Projects requiring quantitative analysis** are projects that have the potential for meaningful differences among project alternatives Freight examples would include development or expansion of a major truck/rail intermodal yard or a major port expansion or access improvement.

Emissions Standards

The Clean Air Act and its amendments also set emissions standards for new engines and vehicles, including freight vehicles. Diesel engines used in freight applications can be grouped into two broad categories:

- **On-road engines,** which include tractor-trailers and other heavy-duty trucks primarily for highway use. The most recent regulations cover model year 2007 and later engines, and include stringent new caps on PM, NO_x, and other pollutants. These are coupled with the new EPA requirement (since 2006) that all on-road diesel fuel be ultra-low sulfur diesel.

- **Off-road engines,** which include mobile non-road diesel engines such as those found in construction equipment, forklifts, farm tractors, and logging equipment. The most recent standards for these engines are being phased in from 2008 to 2015 and vary according to the power output of the engine. However, this group does not include rail locomotives or marine engines, which are regulated separately.

Truck Emissions Standards

Heavy-duty vehicles are defined as vehicles for commercial use that have a gross vehicle weight rating above 8,500 pounds. EPA emissions standards are divided into three groups depending on the date of vehicle manufacture:

- **Model year 1988 to 2003** regulations consisted of gradual phase-in of more stringent requirements for NO_x (from 10.7 grams per brake horsepower-hour in 1988 to 4.0 in 1998) and PM (to 0.10 g/bhp-hr from 0.6).

- **Model year 2004 to 2006** standards (adopted in October 1997) reflect further efforts to reduce NO_x emissions from diesel-powered trucks. These regulations introduced more stringent limits on emissions of hydrocarbons (VOC) and oxides of nitrogen (NO_x); limits on other pollutants like CO and PM remained at 1998 levels.

- **Model year 2007 and later** standards for heavy-duty highway engines were adopted in December 2000 and include regulations covering both engine emissions and diesel fuel. New limits for emissions of hydrocarbons, particulate matter, and NO_x are dramatically lower than previous limits. The new standard for PM emissions took effect in the 2007 model year; new limits on NO_x and hydrocarbons are to be phased in between 2007 and 2010. The fuel requirement, meanwhile, mandated the adoption of ultra-low sulfur diesel fuel for all on-road applications as of mid-2006. This fuel enables the use of advanced pollution control devices such as those discussed in Section 3.0 below.

Locomotive Emissions Standards

EPA regulations governing locomotive emissions are organized into five Tiers (0 through 4). Tiers 3 and 4 were introduced in new EPA regulations in 2008, along with strengthened standards for Tiers 0 through 2. The EPA uses a dual-cycle approach, meaning that all locomotives must comply with both line-haul and switch standards. Tiers 0 through 2 are the currently applicable emissions standards:

- **Tier 0** is the first set of standards, which became effective in 2000 and applies to locomotives and locomotive engines built between 1973 and 2001, any time they are manufactured or remanufactured. EPA's 2008 regulations introduced more stringent requirements for remanufactured equipment, to be phased in by 2010.

- **Tier 1** regulations apply to locomotives and engines originally manufactured between 2002 and 2004; these locomotives are required to meet Tier 1 standards on the date of manufacture and at each subsequent overhaul. As with Tier 0, EPA's 2008 regulations introduced more stringent requirements for remanufactured equipment to be phased in by 2010.

- **Tier 2** standards apply to equipment manufactured in 2005 and later.

They are required to meet the Tier 2 standards at the time of manufacture and each subsequent remanufacture. In addition, as for Tiers 0 and 1, EPA's 2008 regulations introduced more stringent requirements for remanufactured equipment to be phased in by 2010.

- **Tier 3** standards are near-term regulations to be phased in starting in 2011 (for switch locomotives) and 2012 (for line-haul equipment). They will apply to both new and rebuilt locomotives. These standards are to be met through engine technology.

- **Tier 4** regulations are longer term in nature and are expected to be met through the use of exhaust gas after treatments such as those discussed in Section 3.0. The Tier 4 standards are to be phased in starting in 2015 for both switch and line-haul locomotives.

Marine Engine Emissions Standards

As noted in Section 2.2.1, marine diesel engines are divided into three categories. The categories are based on displacement per cylinder.[33] The EPA regulates marine engines differently based on which category they fall into:

- **Category 1 and 2** engines include any that displace less than 30 liters per cylinder and typically range between 700 and 11,000 horsepower. These engines are subject to a tiered system like that for locomotives. Tier 2 standards currently are applicable and govern acceptable emissions of CO, NO_x, and PM. Tier 3 and 4 standards began in 2009 and rely on engine technology and the use of exhaust gas after treatment devices.

- **Category 3** engines are those that displace more than 30 liters per cylinder, and can have power output ranging anywhere from about 3,000 to more than 100,000 horsepower. EPA has adopted NO_x emissions standards for these engines that apply to vessels flagged or registered in the United States equipped with Category 3 engines built in 2004 and later. The limits are the same as those adopted by the International Maritime Organization (a United Nations body) through international negotiation. (As a practical matter, since most vessels calling on U.S. deepwater seaports are registered and flagged in foreign countries, the IMO regulations are the constraining factor.[34]) The residual fuel

---

[33] Cylinder displacement should not be confused with vessel displacement, which refers to the mass of a ship and the equivalent amount of water the vessel displaces while floating.

[34] Some jurisdictions, most notably California, are now requiring vessels to switch to cleaner fuels within a certain distance of shore. These strategies are discussed under *Alternative Fuels* below.

typically used by Category 3 engines limits the emission control technologies that can be used on them, which is why emissions other than NO_x are unregulated.

There are some regional differences in standards for marine engine emissions and fuels. For example, the California Air Resources Board (CARB) is implementing regulations that will require oceangoing vessels operating within 24 miles of the California coastline to use diesel fuel with 0.5 percent or less sulfur content beginning July 1, 2009. By 2012, vessels must use fuel with 0.1 percent or less sulfur.

Non-Road Engine Emissions Standards

There are two types of non-road engines: mobile and stationary. Mobile engines are often found on self-propelled vehicles, but portable equipment (such as generators) are included. Mobile engines are used in a wide variety of applications, including construction vehicles, forklifts, and farm tractors. Mobile cargo handling equipment such as container lifts are a freight-specific example. Stationary engines are not portable and do not appear on self-propelled vehicles; a freight example would be a diesel-powered cargo crane.

Mobile diesel engines are subject to a set of tiered emissions standards similar to those adopted for locomotives:

- **Tier 1 through 3** standards currently are in effect, having been implemented in steps since 1998. The regulations have been phased in for newly manufactured engines in different years depending on the power output of the engine. Each tier represents a progression to more stringent regulations. Sulfur content in non-road diesel fuel was unregulated.

- **Tier 4** standards are to be implemented from 2008 to 2015 and call for stricter limits on NO_x and PM (about a 90 percent reduction in emissions); acceptable CO emissions are unchanged. The EPA also has mandated the use of lower sulfur diesel fuel to enable more advanced pollution control on these engines. Sulfur content for non-road, locomotive, and marine fuels has been limited to 500 parts per million since June of 2007; beginning in 2010, ULSD is required, which only contains 15 parts per million of sulfur.

Emissions from stationary engines were previously unregulated by EPA, which had led to a complex patchwork of state and local regulations. In 2003, Environmental Defense Fund (an advocacy group) brought a lawsuit against the EPA to require the agency to promulgate regulations governing stationary engines. In the consent decree that settled the suit, EPA agreed to adopt emissions standards for stationary engines. As a result, most stationary engines are subject to the same Tier 1 through 4 emissions requirements as mobile non-road engines. However, stationary diesel engines displacing 10 or more liters per cylinder are subject to the Tier 2 standards for Category 2 marine engines.

Aircraft Emissions Standards

The International Civil Aviation Organization (ICAO) typically leads the development of emissions standards for aircraft. ICAO standards cover NO_x and CO emissions, as well as smoke and vented fuel. The latest standards were adopted in 2005 and apply to commercial aircraft engines certified after December 2007 generating more than 26.7 kilonewtons of thrust. The limits are based on a reference landing and takeoff cycle below 3,000 feet, but they also help limit high-altitude emissions (NO_x is a precursor to ozone, which is a greenhouse gas at altitude). Aircraft emissions standards in the United States have been aligned with ICAO standards since 1997. The EPA sets U.S. emissions standards for aircraft, which are enforced by the Federal Aviation Administration (FAA).

At the present time, EPA does not regulate GHG emissions from aircraft, which account for about 10 percent of transportation-related GHG emissions according to the U.S. DOT.[35] However, there is growing pressure for them to do so, as shown by a series of petitions filed by states, regional governments, and environmental groups in late 2007. The petitions requested that the agency make a determination as to whether GHG emissions from aircraft and marine vessels present a danger to public health, and if so, to issue regulations controlling them. The EPA issued an Advanced Notice of Proposed Rulemaking (ANPR) in July 2008. Although the ANPR does not directly address the petitioners' requests, it does compile comments from other agencies on regulating GHG emissions and raises potential issues that may be encountered.

2.3.2 National Environmental Policy Act

The National Environmental Policy Act (NEPA) was signed into law on January 1, 1970. The Act establishes national goals for the protection, maintenance, and enhancement of the environment and stipulates the process for implementing the goals within and among different Federal agencies. It represents a national framework for environmental protection. The basic premise of the law requires the Federal government to create and maintain conditions allowing man and nature to coexist in harmony. Federal agencies are required to incorporate environmental considerations into everyday decision-making using a consistent, systematic approach.

---

[35] U.S. DOT Center for Climate Change and Environmental Forecasting.

Transportation projects (including those for freight transportation) are required to undergo the NEPA process if they involve Federal funds or permits. Air quality is one of the impacts considered in the NEPA process.[36]

The NEPA process varies based on the scope of the proposed project. There are three levels of NEPA analysis:

- **A Categorical Exclusion** is issued when a project is deemed to have no significant environmental impact. Many agencies have lists of actions that are categorically excluded from detailed environmental analysis.

- **An Environmental Assessment (EA)** is prepared if an action is not a categorical exclusion but it is not known whether the action would have a significant impact on the environment. If the EA finds the action would have no significant impact, the agency issues a Finding of No Significant Impact (FONSI).

- **An Environmental Impact Statement (EIS)** must be prepared if it is found that the undertaking will have a significant environmental impact. The EIS is a detailed analysis of the proposed project and its alternatives and solicits public input. The findings of an EIS must be incorporated into an agency's decision-making process.

2.3.3 State Regulations and Responsibilities

Many states have implemented their own air quality and environmental impact laws. These laws are typically very similar to the Clean Air Act and NEPA. Some states enact NEPA-like environmental clearance processes to deal with projects that do not receive Federal funding and thus are not subject to NEPA. Other states have determined that NEPA requirements do not go far enough in terms of environmental protection, and have, therefore, instituted more stringent regulations. The most notable is California, which faces significant air quality challenges due to its high population (and ensuing traffic congestion and growth in vehicle-miles traveled) and its status as a marine freight gateway for the entire United States. Certain geographic characteristics (mountainous areas with valleys in between) combined with climate conditions such as prevailing wind patterns also create areas that are uniquely susceptible to air quality problems.

The California Environmental Quality Act (CEQA) requires all public agencies to "avoid or minimize environmental damage where feasible."[37]

---

[36] 40 CFR 1508.27(b)(2) and (10).

[37] Title 14 California Code of Regulations, Chapter 3, "Guidelines for Implementation of the California Environmental Quality Act."

Structured similarly to NEPA, government agencies in California are required to consider the environmental impacts of public and private activities that they regulate. Like NEPA, there is a list of projects that are exempt from CEQA requirements. Similar to NEPA, although CEQA does not apply to the development of regional or state transportation plans and programs, the projects that are developed following those plans and programs are subject to CEQA requirements. For projects that involve more than one public agency, a designated lead agency prepares the required documentation; the other agencies are required to consider those documents before approving or acting upon the proposed project. In this way, CEQA mandates extensive coordination between public agencies when reviewing projects, including those related to transportation.

2.3.4 LOCAL REGULATIONS AND RESPONSIBILITIES

As mentioned above, MPOs are largely responsible for making determinations of transportation conformity with respect to their planning and programming activities. Local agencies often enact other regulations that can affect freight transportation and emissions. These can include anti-idling laws and other actions designed to minimize the impact of freight movements on local communities. Municipal public works or transportation departments can be instrumental in providing adequate access to freight-generating businesses like distribution centers. Local jurisdictions also are normally responsible for land use and zoning restrictions, which can affect freight movements and freight-related emissions. Land uses have an impact on truck travel patterns (including volume and the way trucks are distributed) since some land uses generate more freight than others. Similarly, localities can collocate industrial uses with intermodal freight facilities to increase the viability of rail, which can affect air quality.

3.0 Strategies for Freight Transportation-Related Emission Reduction/Air Quality Improvement Projects

There are many strategies available to state and local transportation planners and air quality practitioners to reduce emissions from the freight sector. These range from technology applications to operational, policy, and regulatory initiatives. This section describes the different strategies and explains how they work. It also outlines their emissions benefits, cost considerations, possible interactions with other strategies, and co-benefits that may result (such as decreased noise). These strategies can be implemented as standalone projects to improve air quality, or they can be incorporated into transportation projects as environmental mitigation measures. The strategies are divided into two categories:

- **Technology strategies,** which include engine treatments, repowering, alternative fuels, and energy efficiency improvements; and

- **Operational and transportation system management strategies,** which include anti-idling strategies, congestion management techniques, and operational changes that freight generators and private businesses can employ to reduce emissions.

These strategies fall into a wide range of cost, benefit, and timeframe considerations.

3.1 TECHNOLOGY STRATEGIES

Technology strategies to reduce freight emissions take many forms, including retrofitting existing engines with more modern emission control equipment, replacing older engines with cleaner running ones, the use of alternative fuels, and the use of more energy-efficient engines and equipment. Table 3.1 summarizes the most common technology applications for reducing diesel emissions. Several typical applications of each broad strategy type are outlined, along with key issues and considerations for each. Readers who require more information about a particular strategy or family of strategies can consult the detailed descriptions that follow.

Table 3.1 Summary of Technology Strategies to Reduce Freight Emissions

Strategy Type	Purpose	Typical Applications	Key Issues
Exhaust Aftertreatments	Remove pollutants from the exhaust stream	• Diesel Particulate Filters (removes PM, CO, and sometimes NO_x)	• Require ULSD, so applications primarily limited to on-road trucks • Exhaust gases must be held at specified temps for some applications (passive)
		• Diesel Oxidation Catalysts (removes PM and CO)	• Do not require ULSD, making it cost-effective treatment for off-road equipment • Overall emissions reduction limited, compared to other strategies
		• Flow-Through Filters (removes PM, CO)	• Not as effective as DPFs • Can be used by any fuel type, making it useful for older trucks and off-road equipment (including locomotives, marine vessels)
		• Selective Catalytic Reduction (removes NO_x)	• Require tuning, making them better suited for engines with predictable duty cycles (e.g., marine vessels) • Not well-suited for trucks due to varying engine loads
Repowering	Replace older engines with cleaner burning equipment	• New engine/engine (reduces all pollutants)	• New engines have latest emissions control technology • Pre-2007 engines are superior to most older engines, but not as advanced as brand new ones
		• New vehicle replacement (reduces all pollutants)	• Can be more cost-effective to replace old vehicles altogether • Sponsors must ensure old engines are scrapped and do not reenter service
Alternative Fuels	Adopt cleaner-burning fuels	• Liquefied petroleum gas (reduces NO_x, PM, GHG)	• Inappropriate for marine and rail applications due to low-energy content • Fuel distribution network already exists
		• Natural gas (reduces PM)	• Requires special fueling facilities • Similar performance to diesel
		• Biodiesel (reduces PM, CO)	• 20 percent biodiesel blend can be used without engine modifications • May slightly increase NO_x emissions
		• Fuel-borne catalyst (reduces PM)	• EPA cautious about their use as may increase emissions of some particles
		• Low-sulfur diesel (reduces PM), emulsified diesel (reduces PM, NO_x)	• No engine modifications required, but more expensive fuel • Emulsified fuel contains less energy per gallon than conventional diesel
		• Fuel cells	• Not practical at this time
Energy Efficiency	Save fuel/emissions through superior design	• Hybrid-electric vehicles	• Available for medium-duty tractor trailers • Most suited for trucks operating primarily in urban areas
		• Weight-saving modifications (reduce all pollutants)	• Often low-cost • Enhances productivity by increasing payload capacity
		• Improved aerodynamics, reduced rolling resistance (reduce all pollutants)	• Fuel savings may offset capital outlay, making this a cost-neutral strategy
		• Marine vessel efficiency improvements	• Generally not controlled by ports
		• 'Green' locomotives (reduce all pollutants)	• Genset or similar rail equipment can be expensive • These investments be combined with larger capital improvements

3.1.1 Aftertreatment ("Tailpipe")/Engine Controls

This category includes emission control devices that can be integrated into both new engines and retrofits. This handbook will focus on retrofits, since new engine standards are largely addressed by Federal regulations for engine manufacturers. There are several types of retrofits available for freight diesel engines:

- **Diesel Particulate Filters (DPF)** – These devices remove particulate matter from diesel exhaust. DPFs used in freight vehicle applications typically dispose of accumulated particles by burning them off in a process known as "filter regeneration." Sulfur in diesel fuel can interfere with filter regeneration, which is why the use of ULSD is necessary to achieve maximum emission reductions with this technology. In combination, a DPF installed on an engine using ULSD can reduce PM and CO emissions by 60 to 90 percent.[38] Figure 3.1 shows how the technology works. Diesel exhaust enters the flow channels in the filter (Step 1), but is blocked by walls at the end of the channels (Step 2). This forces the exhaust gases to move through the porous walls of the filter. Particulate matter is captured on the walls and burned off during filter regeneration. DPFs are best suited to truck applications because they require low-sulfur fuel. However, they have been successfully used in locomotives (the BNSF and UP railroads have both retrofitted some of their locomotives with DPFs). The high sulfur content of bunker fuel makes DPF technology impractical for marine cargo applications.

Figure 3.1   Diesel Particulate Filter

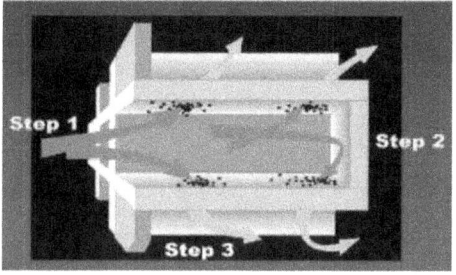

Source: Environmental Protection Agency.

Diesel particulate filters are divided into two categories, depending on how filter regeneration is accomplished:

- **Passive DPF** uses a catalytic material which enables trapped particulate matter to be burned off at a lower temperature. Exhaust gases must be at a specified temperature for a certain period of time in order for this technology to work; otherwise, the filter will become plugged with soot, interfering with filtration and eventually causing engine damage. It is therefore important to verify the duty cycle and operating characteristics of equipment proposed for retrofits with this technology.

- **Active DPF** systems do not use exhaust gas heat to burn off trapped PM. Instead, they regenerate by passing electrical current through the filter, adding fuel to achieve the necessary combustion temperature, or adding a catalyst that reacts with the PM. Active DPF can be used in engines with lower exhaust gas temperatures.

---

[38] http://www.epa.gov/oms/schoolbus/retrofit.htm.

- **DPF with NO_x Catalyst –** Some DPF technologies have been coupled with NO_x catalysts to control NO_x emissions. A NO_x catalyst is installed downstream from the DPF; since the particulate matter already is removed from the exhaust gases, the catalyst can work without getting clogged up by the soot.

- **Diesel Oxidation Catalyst (DOC) –** Diesel oxidation catalysts use a chemical process to break down the pollutants found in diesel exhaust, converting them into less harmful compounds. These devices can reduce PM emissions by 20 percent and CO pollutants by up to 40 percent. Unlike DPFs, DOCs do not require the use of low-sulfur fuel. DOC technology only works on the soluble organic fraction of diesel particulate matter emissions, which is why the overall emissions reduction is limited. DOCs are suitable for truck and rail applications as well as some marine applications, but the technology is not yet fully developed for the largest marine engines.

- **Flow-Through Filter (FTF) –** Flow-through filters work by forcing exhaust gases to flow through a filter material (such as wire mesh) that introduces turbulence to the exhaust flow. This medium is treated with a catalyst that reduces emissions of PM and CO. Because the exhaust is interrupted as it passes through the filter, it spends more time in contact with the catalyst, thereby reducing emissions. FTFs are not as effective as DPFs at removing pollutants, but they also are less likely to become clogged up and can be used with any type of diesel fuel. This makes them ideal for engines or operating environments that may be unsuitable for DPF applications, such as trucks using off-road diesel fuel (e.g., logging trucks), locomotives, and cargo ships.

- **Selective Catalytic Reduction (SCR) –** SCR is a technology for controlling NO_x emissions that uses a catalyst to convert NO_x to nitrogen and water. Installed downstream of a DPF, the SCR system injects diesel exhaust fluid[39] into the hot exhaust gases, which then travel through a catalyst where they are converted to nitrogen and water and emitted through the tailpipe. Although these systems are most frequently found in industrial applications such as utility boilers, they have successfully been applied to marine diesel engines, locomotives, and even automobiles. The chief obstacle to adopting this technology for a wider range of vehicles is the need to tune the SCR system to the operating cycle of the engine. Engines with predictable duty cycles

---

[39] Diesel exhaust fluid is a solution of water and urea (an organic compound also known as carbamide).

(such as large cargo ships) are well suited to SCR retrofits.[40] Trucks' operating cycles vary widely depending on many factors like driver habits, stop-and-go traffic, and hilly terrain, making it harder to use SCR systems.

3.1.2 REPOWERING

Repowering refers to replacing an old engine with a newer, cleaner engine or converting to electric power (as in certain types of cargo-handling equipment, such as port gantries). Government agencies often offer tax credits or other incentives to encourage businesses to repower equipment. In general, there are four options for repowering:

- **New Engine –** The old engine is replaced with a brand new one that meets all of the latest emissions control regulations;

- **Older (pre-2007) Engine –** The old engine is replaced with an engine manufactured before 2007 and retrofitted with an emissions control device such as those discussed above;

- **Alternate Fuel/Electricity –** This option involves converting the equipment to run off electricity or an alternate fuel, such as propane. This is a common practice for cargo-handling equipment such as cranes and forklifts; and

- **New Vehicle Replacement –** In some cases, it can be more economical to simply replace a piece of equipment with an entirely new model that employs the latest emissions control technology. In those instances, agencies can offer incentives to get businesses to replace older equipment, thereby removing more polluting vehicles from service.

No matter which strategy is employed, it is important to ensure that the old equipment is scrapped rather than sold and put back into service. This will ensure that the emissions benefit is fully realized.

3.1.3 ALTERNATIVE FUELS

There are several alternative fuel technologies available that provide cleaner-burning options for freight vehicles and equipment. The focus here is on common alternative fuels using proven technologies that already are available; potential advanced technologies that are not yet practical (such as fuel cells) are discussed briefly to make readers aware of their current status.

---

[40] There have been successful demonstrations of stationary SCR systems at rail yards that capture exhaust gases with a fume hood positioned above the railroad tracks and transfer it to an emissions treatment system.

• **Natural Gas** – Compressed natural gas (CNG) is a mixture of hydro-carbons (primarily methane, a greenhouse gas) extracted from gas wells or produced in conjunction with crude oil. Vehicles powered by CNG perform similarly to those powered by diesel fuel, but CNG vehicles emit 70 to 90 percent less particulate matter than conventional diesels, since burning natural gas produces virtually no particulate matter. CNG may offer limited GHG reduction benefits, although one study of heavy-duty applications found that on a life-cycle basis GHG emissions were approximately equal to those from diesel fuel.[41] CNG fleets require special refueling and maintenance facilities due to the specific requirements for handling and storing CNG. CNG fueling infrastructure can be found all over the country, but is some-what sparse in some Rocky Mountain states, the Great Plains, and the South (see Figure 3.2). California and certain parts of New England have the most CNG stations. CNG has an added advantage in that the majority of it is produced in the United States, reducing the nation's dependence on foreign oil. CNG-powered equipment costs significantly more than equivalent diesel-powered vehicles. The San Pedro Bay Ports in Southern California are now running a demon-stration project with CNG-powered trucks that transport containers from ships to consolidation yards in the area. The project is part of the San Pedro Bay Ports Clean Air Action Plan, which is detailed in Section 5.1.3. A variation is liquefied natural gas (LNG), which is a better choice for rail applications because of its greater density, which reduces the frequency of refueling.

Figure 3.2   Natural Gas Fueling Stations by State
2009

Source: U.S. Department of Energy Alternative Fuels and Advanced Vehicles
          Data Center.

---

[41] Clark, W., 2007: "Market Penetration Issues for Biodiesel." National Renewable Energy Laboratory, http://www.sae.org/events/sfl/presentations/2007clark.pdf.

- **Liquefied Petroleum Gas (LPG)** – Commonly known as propane, LPG can be used to replace gasoline in light-duty vehicles and diesel in heavy-duty vehicles. Several original equipment manufacturer (OEM) LPG engines are available for heavy-duty vehicle use, including delivery trucks, school and shuttle buses, and recycling trucks. In addition, a well-developed distribution network of LPG fueling stations already exists. However, LPG has a lower energy content than traditional fossil fuels, making it inappropriate for marine and rail freight vehicles because of the increased refueling frequency.

- **Low-Sulfur Fuels for Marine Engines while in Proximity to Shore** – As discussed in Section 2.2, most large ocean-going vessels utilize bunker fuel, which is inexpensive but has high levels of sulfur. In response, many coastal jurisdictions are now mandating the use of low-sulfur diesel fuel by cargo vessels when they are operating close to shore. California is a pioneer in this area; in July 2008, the State began requiring large vessels to use low-sulfur fuel whenever they are within 24 miles of the coast. The new rules require ships to burn fuel with 0.5 percent or less sulfur in coastal waters beginning in 2009; in 2012, they must use fuel with 0.1 percent or less sulfur content. By contrast, bunker fuel typically has a sulfur content of 3.5 percent. Complying with these rules requires that ships have the capability to be "dual-fueled"; that is, they must be designed or retrofitted with a separate fueling system allowing the use of distillate fuel in the auxiliary engines. Research has shown that many large vessels such as containerships already have separate fuel tanks for their auxiliary engines. These vessels have the potential for dual-fuel operations, but currently operate on residual fuel when on-shore due to its lower cost – fuel costs are a major component of ocean shipping. Vessel manufacturers have been responding by incorporating more low-sulfur tanks in new ships.[42]

- **Emulsified Diesel Fuel** – Emulsified diesel fuel is a mixture of diesel fuel with water and emulsifying additives. This reduces PM and NO_x emissions, but emulsified fuel also contains less energy due to the addition of water, resulting in power losses and decreased fuel economy. The presence of water also can be problematic; if a vehicle sits unused for too long, the water will separate from the fuel, which can harm the engine. Emulsified diesel also costs about 20 cents more per gallon than regular diesel. EPA data show that the use of this fuel can reduce PM emissions by 20 to 50 percent and NO_x by 5 to 30 percent.[43]

---

[42] Port of Los Angeles and Starcrest Consulting Group LLC, *Evaluation of Low Sulfur Marine Fuel Availability*, July 2005.

[43] United States Environmental Protection Agency, Office of Transportation and Air Quality, *Clean Fuel Options for Heavy-Duty Trucks and Buses*, June 2003.

- **Ultra-Low Sulfur Diesel Fuel –** As mentioned in Section 2.2, the EPA has mandated the use of ULSD in new on-road trucks since the 2007 model year, and this fuel has been available at retail stations since mid-2006. ULSD (defined as diesel fuel with 15 parts per million or less sulfur content) enables the use of more advanced emissions control technologies such as the diesel particulate filters described above. Locomotives, marine engines, and off-road trucks are not required to use ULSD at this time, so there is an opportunity to further reduce emissions by adopting low-sulfur diesel fuel in these other modes.[44]

- **Biofuels –** This is a broad category of alternative fuels that includes gasoline substitutes such as corn or cellulosic ethanol. The most relevant for freight transport is biodiesel, which is a renewable fuel that can be produced from vegetable oils and animal fats. Biodiesel is safe and biodegradable. It can reduce PM, CO, and hydrocarbon (HC) emissions, but may also slightly increase NO_x emissions. A blend of 20 percent biodiesel and 80 percent conventional diesel (known as B20) can be used in diesel engines without requiring modification. This blend reduces PM emissions by about 10 percent, but increases NO_x emissions by two percent. Pure biodiesel (B100) reduces PM emissions by about 40 percent, but often requires engine modifications to work and is typically not suitable for cold climates.[45]

- **Fuel-Borne Catalysts (FBC) –** Also known as fuel additives, FBCs are metallic chemicals added to diesel fuel to improve combustion and thereby reduce PM emissions. These additives can reduce oxidation temperatures for PM, so that a DPF would not have to reach as high a temperature to enable soot in the exhaust to be burned off. However, it should be noted that these additives, when used in dosages above a certain level, can increase emissions of very fine metal oxide particles. For this reason, the EPA has been cautious about the use of FBCs. To minimize the amount of metals discharged into the atmosphere while maximizing emissions reductions, FBCs can be combined with retrofits such as a DPF.[46]

- **Fuel Cells –** Fuel cells are an emerging technology that may have useful applications for transportation in the future. Unlike hybrid-electric vehicles, which store energy from an external source in an on-board battery, fuel cell vehicles (FCV) create their own electricity. This is accomplished through a chemical reaction using hydrogen

---

[44] See, for example, the new California regulations requiring low sulfur fuels for oceangoing vessels operating near the coastline discussed in Section 2.3.2.

[45] Ibid.

[46] Environmental Defense, "Cleaner Diesel Handbook: Bring Cleaner Fuel and Diesel Retrofits into Your Neighborhood," April 2005.

fuel and oxygen from the air. The hydrogen can be supplied either as pure hydrogen stored in a tank or from hydrogen-rich fuels such as methanol, natural gas, or even gasoline. The latter technique requires a "reformer" to extract pure hydrogen from the fuel for use in the fuel cell; this process emits some carbon dioxide, but not nearly as much as a conventional engine. Vehicles fueled by pure hydrogen emit no pollutants; the only byproducts of power generation are water and heat. There are a number of obstacles to the widespread adoption of this technology for the transport sector, including cost, distribution, and storage of hydrogen fuel, and competition with other technologies such as hybrids. Although limited quantities of FCVs, such as the Honda FCX Clarity, are available to the public, they will not be mass-produced for a number of years, and freight fuel-cell vehicles may take longer to be commercialized.

3.1.4 ENERGY EFFICIENCY

More efficient engines and equipment generally reduce emissions of all pollutants, including GHG. This includes a variety of options such as hybrid-electric vehicles, improved vehicle aerodynamics, more efficient tires, and reduced vehicle weight. Energy efficiency strategies have the advantage of reducing fuel costs, sometimes making them cost-neutral. EPA's SmartWay transportation program (discussed in more detail in Section 4.4), which packages a suite of efficiency improvements specifically for truckers, is premised on the notion that the package pays for itself. Most of the truck upgrades identified below qualify for special low-interest financing through the SmartWay transportation program.

- **Hybrid-Electric Vehicles** – Hybrid-electric technology, already becoming widespread in passenger vehicles, is now available for medium-duty tractor-trailers. There are now diesel hybrid-electric tractors available targeted towards general freight haulers and food/beverage distributors. These tractors use an electric motor with an automated transmission/clutch combined with a traditional internal combustion diesel engine and transmission, and utilize regenerative braking to recapture power otherwise lost during deceleration and braking.[47] Regenerative braking is more effective in large commercial vehicles because their greater mass requires more power to stop, meaning there is more potential energy to capture and reuse. This technology can cut fuel consumption by 25 to 50 percent depending on the application. Although hybrid-electric vehicles can cost up to

Figure 3.3 Typical Hybrid-Electric Propulsion System

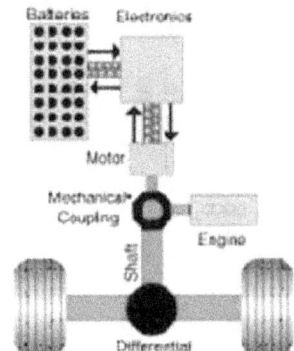

Source: Electric Transit Vehicle Institute.

---

[47] Regenerative braking captures the kinetic energy of the vehicle (which would otherwise be wasted as heat through conventional braking) and stores it in the vehicle's battery to provide motive power. It is distinct from the dynamic braking used by trains.

$40,000 more than regular trucks, a Federal tax credit, the Alternative Motor Vehicle Credit, is available to offset this cost. It contains provisions for Qualifying Heavy Hybrid vehicles, which are defined as new vehicles with a gross vehicle weight over 8,500 pounds that meet the definition of a qualifying hybrid vehicle; the Internal Revenue Service maintains a list of such vehicles.[48] Hybrid vehicles are most effective in stop-and-go traffic, suggesting that hybrid vehicles are best suited for urban applications such as delivery trucks. Figure 3.3 shows how a typical hybrid-electric truck system works. The electric motor supplies additional power from a battery pack to supplement the diesel engine and while recharging the batteries through regenerative braking.

Figure 3.4   Example of Freight Vehicle Aerodynamic Improvements

Source: Environmental Protection Agency.

- **Improved Vehicle Aerodynamics –** At highway speeds, wind resistance (aerodynamic drag) accounts for the preponderance of truck energy losses. Similarly, line-haul freight trains lose a significant amount of energy to drag because of their aerodynamically unfavorable profile, unshielded space between cars, and lack of covers on empty cars. Improving vehicle aerodynamics is another way to cut fuel consumption and emissions. Aftermarket fairings attached to the front and/or belly of truck trailers can improve fuel efficiency by up to six percent (Figure 3.4). There also are modifications to the tractor that can improve fuel economy, such as upgraded front bumpers, air dams, and side mirrors. Roof fairings, cab extenders, and side fairings installed on a tractor can achieve fuel savings of up to 600 gallons per year and emissions reductions of over five metric tons of GHG. For instance, in Figure 3.5 the air dam visible above the truck cab improves air flow and increases fuel efficiency. Similarly, covering empty rail cars, modifying how intermodal cars are loaded, and minimizing open areas between cars can help improve train aerodynamics. It also is relatively inexpensive to implement these strategies, either by ordering them as options on new trucks or rail cars or by retrofitting older equipment, so the costs are recouped quickly through improved fuel efficiency.[49]

- **More Efficient Tires –** Tire rolling resistance accounts for about 13 percent of truck energy use. Many truck fleets have begun to switch to more fuel-efficient single wide tires, which replace the traditional dually style tires found on most tractor-trailers. These tires improve fuel efficiency by reducing weight and rolling resistance (although impacts to infrastructure are not yet quantified). The EPA has found that the use of single wide-base tires can improve fuel economy by

---

[48] http://www.irs.gov/businesses/article/0,,id=175456,00.html.

[49] U.S. Environmental Protection Agency SmartWay Transport Partnership, "A Glance at Clean Freight Strategies: Improved Aerodynamics," February 2004.

two to five percent over conventional dual tire setups. On a combination long-haul truck, this equates to a fuel savings of up to 400 gallons per year, and a reduction in CO emissions of four metric tons annually.[50] In addition, wheels for these tires cost less than dual wheels, while the tires themselves are cost-competitive with equivalent dual tires. This makes single wide tires an attractive option for new trucks.

- **Reduced Wheel-to-Rail Friction –** Railroads periodically apply grease to their tracks to reduce fuel consumption and protect infrastructure from excessive wear. Conventional lubrication systems apply large and uneven amounts of lubricant to the rail, resulting in wasted material and poor transfer of grease to the passing train wheels. Newly developed computer controlled systems provide a better application of lubricant and limit the amount of grease applied to reduce excessive applications that lengthen required braking distances. Wheel and rail wear is reduced as well as fuel consumption.

- **Weight Reduction –** Reducing the weight of a freight vehicle directly affects the energy required to move it and, therefore, has an impact on emissions. Owners can install weight-saving devices on truck tractors such as aluminum alloy wheels and aluminum axle hubs that replace heavier steel components. There are even greater opportunities to save weight in the trailer, where aluminum parts can be used in the roof and upright posts as well as floor joists. Overall, these modifications can reduce the empty truck weight (known as "tare weight") by up to 3,000 pounds, saving between 200 and 500 gallons of fuel annually and reducing greenhouse gas emissions by two to five metric tons per year. These weight restrictions also allow increased payload which may result in no change in gross vehicle weight and fuel savings per mile, but may reduce the miles travel and increase productivity of the vehicle. Aluminum rail cars already are available and in use, and are up to one-third lighter than comparable steel cars. Aluminum also offers exceptional resistance to corrosion from certain cargoes (like high-sulfur coal) and is more valuable for recycling purposes when the rail car is scrapped. Modifications that reduce weight have the added benefit of allowing the vehicle to carry a larger payload, thereby improving productivity. Lighter-weight trucks and trailers do command a price premium since the lightweight components are more expensive, making them more common in weight-sensitive applications like heavy goods and refrigerated foods.[51] The higher carrying capacity of aluminum rail cars may recoup their greater initial cost within two years.

---

[50] U.S. Environmental Protection Agency SmartWay Transport Partnership, "A Glance at Clean Freight Strategies: Single Wide-Based Tires," February 2004.

[51] U.S. Environmental Protection Agency SmartWay Transport Partnership, "A Glance at Clean Freight Strategies: Weight Reduction," February 2004.

- **Marine Vessel Efficiency Improvements** – There also are a number of improvements that can be made to cargo ships to make them more fuel efficient. Although port planners and other officials typically have little control over things like ship design, some of the available technologies are summarized here for readers wishing to explore these options:

  - **Energy-efficient paint for vessel hulls.** Ship owners must regularly paint their hulls to avoid fouling by barnacles and other marine life. There are a variety of hull paints that reduce drag and fuel consumption. The Emma Maersk, one of the largest containerships in the world, uses a silicone-based hull paint that improves efficiency by creating a very slick surface that reduces drag and helps to prevent fouling. It should be noted that many marine paints contain toxicants that are released over time, especially during underwater hull cleaning, thereby contributing to water pollution problems. However, there are paint choices (such as the silicone paint) that contain no biocides.

  - **Exhaust heat recovery systems.** Another technology application available for cargo ships is exhaust heat recovery. These systems pass hot exhaust gases through a steam generator, which powers electrical generators to generate electricity for shipboard use. On the Emma Maersk, such a system produces electrical power equivalent to about 12 percent of engine output while also using the steam to provide heat.

  - **Improved hull design.** Certain hull designs offer better hydrodynamics than others. For example, reducing vessel displacement by increasing the hull width by only 0.25 meters allows for a reduction of 3,000 tons of ballast, reducing propulsion energy requirements by 8.5 percent.[52] Use of interceptor or trim planes (vertical plates fitted to the ship's transom) can improve fuel consumption by 1 to 4 percent, or up to 10 percent if the interceptor or trim plane is used in conjunction with a ducktail (an extension of the rear of the ship that reduces water resistance).

- Enhanced Locomotive Engine Technologies – There are a number of technologies either existing or under development that promote better fuel efficiency in railroad locomotives, including:

  - Common rail fuel injection systems allow for a more controlled fuel injection rate across all engine speeds by storing fuel at high pressures along a common rail connected to each cylinder. This yields more efficient combustion while providing smoother, quieter

---

[52] Wartsila, 2009.

running engines, reducing GHG emissions and fuel consumption by at least 10 percent.[53] The largest locomotive with a fuel injection engine currently available has a maximum horsepower of 4,000, appropriate for use in yard and line-haul operations.

– Genset yard locomotives use multiple smaller (approximately 700 horsepower) diesel engines to provide only the power that is needed and have electronic engine controls to better match locomotive activities to operating conditions. Older locomotives can be retrofitted with genset engines, which are newer and more efficient than larger conventional yard engines, and are certified to EPA Tier III emission standards. This technology can save between 15 and 24 gallons of diesel fuel per locomotive, per day, with accompanying emissions reductions.

– Hybrid propulsion systems employ a small, efficient diesel engine to charge a set of batteries which provides power to the locomotive, similar to a hybrid system on a passenger car or truck. The engine operates only when the batteries need to be recharged. As a result, the diesel engine can stay within its optimal load range, reducing emissions and fuel consumption. These systems are most effective for switch locomotives in stop-and-go railyard operations, although line-haul versions also have been developed.

The chief limitations of these approaches are their very high capital cost and slow introduction due to the long life cycle of the rail fleet. Some states (like California and Texas) have adopted programs to subsidize the implementation of these technologies. Such programs can help accelerate the adoption of these strategies, providing public benefits before the railroads would do so on their own.

3.2 OPERATIONAL STRATEGIES AND TRANSPORTATION SYSTEM MANAGEMENT

There are a variety of operational and system management strategies that policy-makers can employ to reduce freight vehicle emissions. These usually take the form of local regulations and ordinances (such as anti-idling programs), congestion mitigation efforts geared towards speeding the flow of freight (such as improved port access), or operational changes to reduce emissions (such as speed reduction). Table 3.2 summarizes some of the most common approaches and the potential issues, advantages, and co-benefits of each.

---

[53] International Union of Railways, 2002.

Table 3.2   Summary of Operational and Transportation System Management Strategies

Strategy Type	Purpose	Typical Applications	Key Issues
Anti-idling	Reduce/ unnecessary idling	Truck stop electrification, auxiliary power units (reduces all pollutants)	• Difficult to electrify all potential truck parking locations (rest areas, parking lots) • Auxiliary power units are significant initial cost to truck owner • Fuel savings offset costs
		Anti-idling regulations (reduces all pollutants)	• May be difficult to enforce • Can be written specifically for freight vehicles
		Locomotive idling limit device	• Most effective in warm climates
		Shore power (reduces all pollutants)	• Requires large capital outlay (vessel retrofits and land-side hookups) • Currently no universal standard
Congestion Mitigation	Relieve bottlenecks with capacity improvements or better system management	Port access improvements, truck-only lanes (reduces all pollutants)	• Capacity expansion is typically expensive and has other environmental impacts • Few examples of truck-only lanes in the U.S.
		Signal coordination (reduces all pollutants)	• Low cost, but limited benefits • Some signal systems can detect trucks in traffic mix and adjust timing accordingly
		Rail infrastructure improvements, grade separation (reduces all pollutants)	• High capital costs • Grade separation also improves safety
		Short-sea shipping (reduces all pollutants)	• Time considerations, terminal and drayage costs, and requirements to purchase comparatively expensive U.S. built vessels (Jones Act) make competition with land modes difficult
Operational Changes	Modify business practices to minimize emissions	Speed reduction programs, driver training (reduces all pollutants)	• Voluntary programs may have limited effectiveness • Trucks can be ordered with speed limiting devices • Fuel and maintenance savings may offset productivity loss
		Reduced pickup/dropoff idling for trucks (reduces all pollutants)	• Often implemented as improved port operational strategies
		Weigh station bypass (reduces all pollutants)	• Can improve monitoring and compliance
		Reduced empty mileage, circuitous routing (reduces all pollutants)	• Not all origin-destination pairs have "balanced loads," making backhauls difficult to identify • More difficult to reduce empty rail backhauls • Reducing circuitous rail routing requires rail capacity expansion

3.2.1 ANTI-IDLING

Anti-idling strategies refer to efforts to reduce emissions by cutting down on the time freight vehicles spend idling (sitting in one place with the engine running). These strategies exist for all modes, including trucks, locomotives, and marine cargo vessels, and can be implemented through regulations, technology applications, or a combination of the two. Several strategies are outlined below.

- **Shore Power ("Cold Ironing") –** Cargo ships usually switch to their auxiliary engines to provide power for ship operations while they are in port. Although some auxiliary engines use cleaner distillate fuel than the main engines, they still contribute to localized air pollution around port complexes since the ships may be idling for days at a time. To combat this problem, many ports are constructing shore power (also known as cold ironing) systems that provide clean electrical power to cargo vessels while they are in port. The U.S. Navy has been using cold ironing for decades, not because it cuts emissions, but rather because it reduces equipment wear and tear and saves fuel (which are ancillary benefits of implementing a shore power strategy). California, in general and the Ports of Los Angeles and Long Beach in particular, are driving the development of this technology in the United States. Cold ironing is a key part of the San Pedro Bay Ports Clean Air Action Plan (described below in Section 5.1.3.) The plan calls for all major cargo terminals at the ports to be equipped with shore power by 2016. In addition, the California Air Resources Board (CARB) is requiring all container, passenger, and refrigerated cargo ships to shut off their auxiliary engines while in port. Containerships lend themselves well to cold ironing, since they rely on land side cranes to load and unload cargo, rather than shipboard equipment that must be powered from the ship. Although shore power is a promising way to reduce cargo vessel emissions, capital costs and lack of standards are limiting its widespread adoption.

  - **Significant capital costs.** Cold ironing requires a large up-front capital investment both for ships and landside hookups. Retrofitting a containership typically costs between $200,000 and $500,000. However, some vessels are now being constructed with built-in shore power capability. Ports also can offer incentives to shipping lines to encourage them to adopt the technology; the Port of Long Beach recently signed an agreement with Matson Shipping Company whereby the company will retrofit five of its ships with shore power systems in return for tax incentives and discounted tariffs.

– **No universal standard.** At this time, there is no internationally accepted universal standard for shore power systems. So, a cargo ship that is equipped for cold ironing at one port may not be able to hook up to a system at another port.

• **Idling Limit Devices on Rail Locomotives** – These devices automatically shut off a rail locomotive's engine if it sits idle for a certain period of time, usually 15 minutes. This prevents unnecessary emissions from locomotives that are not in use. As with shore power, the San Pedro Bay ports are leaders in this area. By the end of 2008, all switch/helper locomotives operated by the Pacific Harbor Line (which provides switching services to the ports) were required to be equipped with 15-minute idle limit devices, followed by Class I switchers (in 2011) and line-haul locomotives (in 2014).[54] CARB has promulgated similar regulations for locomotives that operate primarily in the State of California. Governmental agencies sometimes offer grants or other assistance to encourage railroads to install these devices. For example, the Texas Commission on Environmental Quality (TCEQ) offers grants of up to $5,000 per ton of NO_x emissions reduced in certain eligible Texas counties through the use of idle-reduction technology, including idle-limiting devices.[55] These devices achieve their maximum benefit in warm climates, since locomotives typically must be kept running in cold weather to keep the engine from freezing up. However, locomotives can be fitted with systems that monitor key operating parameters (such as engine coolant temperature) in cold weather, and automatically restart the engine as needed.

• **Truck Stop Electrification** – Truck stop electrification (TSE) is basically akin to shore power for trucks. Since Federal Motor Carrier Safety Administration and state agencies limit the number of consecutive hours drivers may operate commercial motor vehicles and/or be on duty, drivers often rest at truck stops or rest areas.[56] Truckers have historically left their vehicles idling during these rest stops for comfort purposes (air conditioning and heat), but idling a truck engine burns almost one gallon of fuel per hour. TSE systems allow truck drivers to instead use electric power for in-cab heating, air conditioning, and other functions. A number of private companies offer TSE technology for fleet owners, independent truck owner-operators, and travel plazas. The systems can be as simple as an extension cord

---

[54] Ports of Los Angeles and Long Beach, "San Pedro Bay Ports Clean Air Action Plan Source Specific Standards."

[55] Texas Commission on Environmental Quality, "Emissions Reduction Incentive Grant: Supplemental Activity Application Forms – Locomotive."

[56] Federal regulations require truckers to take 10 hours of rest for every 11 hours on the road.

hookup running from a parking space power station into the truck cab. Other systems use a window hookup to provide climate control, power outlets, Internet access, and other amenities. Some trucks are now being built with TSE hookups; others can be retrofitted to accept the appropriate connections. One limitation of this strategy is that there is limited parking at truck stops so many truckers rest at decentralized locations such as rest areas or parking lots, which cannot be efficiently electrified. As of October 2008, less than three percent of the nation's 5,000 truck stops were electrified.[57]

- **Auxiliary Power Units (APU)** – Auxiliary power units are devices that are typically installed on trucks to power accessories and climate control systems while the truck is parked, without idling the main engine. They can be battery-powered, but most APU are small diesel generators. Since the APU is considerably smaller than the truck's engine, it has much lower emissions. APU also save fuel costs and reduce unnecessary engine wear. This technology requires a significant up-front cost on the part of the truck owner (the average is about $6,000 per truck); as a result, it is most likely to be adopted on newer trucks with sleeper cabs. Use of the technology will, therefore, likely expand as the nation's truck fleet turns over.

- **Regulations Prohibiting Excessive Idling** – As of 2006, 14 states plus the District of Columbia, as well as many counties and municipalities, had anti-idling regulations.[58] Usually, regulations make it illegal to idle a vehicle beyond a certain length of time, but there are normally exceptions for emergency vehicles, inclement weather, or other factors. These regulations are not always limited to commercial vehicles and also may apply to passenger vehicles. Typically regulations are written specifically for vehicles over a certain weight rating.

3.2.2 CONGESTION MITIGATION

Traffic congestion increases transportation sector emissions because vehicles idling in traffic emit more than those traveling at a steady speed. Efforts to alleviate congestion, therefore, can have a positive impact on emissions, including those from commercial vehicles. This section will focus on efforts specifically targeted towards reducing congestion associated with freight vehicles, but many of these improvements also benefit the traveling public.

---

[57] U.S. Department of Energy, Energy Efficiency, and Renewable Energy Information Center.

[58] U.S. Environmental Protection Agency. *Compilation of State, County, and Local Anti-Idling Regulations.* April 2006.

- **Arterial Signal Coordination on Routes with High Truck Traffic –** Adjusting signal timing to optimize traffic flow on routes with a high percentage of trucks is one way to help minimize freight emissions. A truck traveling at 55 mph that must stop at an intersection and then reaccelerate loses 60 to 80 seconds, as do passenger cars traveling behind the truck.[59] Time spent idling at a traffic signal increases emissions, and reaccelerating increases them even more because the engine must work harder. Therefore, on certain arterial routes that experience heavy truck traffic, it can be beneficial to ensure efficient signal coordination to better facilitate traffic flow. There are signal systems available that can detect the presence of trucks in the traffic mix and adjust signal timing accordingly.

- **Port Access Improvements –** Improving capacity at key access routes to seaports (either road or rail) can reduce emissions by minimizing wait times and queuing at the gates. Although these projects are rarely justified solely by their air quality benefits, the emissions reductions from potential access improvements can be quantified as additional benefits that can help move a project forward. The SR 519 Intermodal Access Project in Seattle (detailed in Section 5.2.2) was found to have numerous air quality benefits associated with reduced idling times for trucks and trains accessing the port.

- **Grade Separations for Road and Rail –** Grade separation refers to physically separating two or more transport corridors to eliminate conflicts between traffic traveling in different directions. It most often refers to the separation of railroad tracks that cross highways, but also can be applied to rail/rail crossings. Highway/rail grade separations can limit emissions from cars and trucks that must stop and wait for trains to pass. They can, therefore, reduce vehicle emissions that are attributable to freight movements while mitigating passenger vehicle congestion at the same time. A rail/rail grade separation (such as the proposed Colton Crossing project in California, discussed in Section 5.2) directly reduces locomotive emissions since trains no longer have to stop for other trains going through the intersection.

- **Rail Infrastructure Improvements –** Making improvements to rail line capacity and infrastructure can reduce freight rail emissions on corridors with heavy train traffic. Upgrading a single track corridor to double track, for instance, eliminates the need for one train to stop at a siding to allow another train to pass. Similarly, improving a rail tunnel to allow double-stacked container cars increases the volume of freight that can be moved by one train, thereby allowing

---

[59] Eyler, D. "Traffic Responsive Signal Coordination." Presentation given to the TRB Traffic Signal Systems Committee, July 2003.

the railroad to reduce the number of trains it operates or, conversely, increase the amount of freight hauled without increasing the number of trains. This is especially important since intermodal rail traffic has been growing much faster than traditional carload traffic over the past several decades. The Heartland Corridor Clearance project is one example of rail capacity improvements that have an ancillary air quality benefit. The project involves track and tunnel modifications that will allow double-stacked container trains to travel between Hampton Roads, Virginia and Columbus, Ohio on a Norfolk Southern rail corridor. Once complete in mid-2010, the upgrades will improve air quality by allowing greater cargo volume on the same number of trains while eliminating the current circuitous route that double-stacked trains must take to get between these points. The project is being funded through a public/private partnership (PPP) between Norfolk Southern, the U.S. DOT, the Virginia Department of Rail and Public Transportation, and the Ohio Rail Development Commission.[60]

- **Truck-Only Lanes –** A few states have experimented with freeway lanes wholly or partially devoted to trucks. By separating trucks from other traffic, truck-only lanes can improve traffic flow (which reduces emissions) and enhance safety. In the United States, this technique is used most often on short highway segments in dense urban areas that have a lot of truck traffic, or that link a port to the regional/national highway system. Though this technique is not widespread in the United States, there are a few examples of truck-only lanes. California was an early adopter of the strategy, and there are now two truck-only lanes (on I-5 in Los Angeles and Kern counties) in operation with more under consideration. Trucks are required to travel in the truck-only lanes, which are marked with black and white signs; automobiles are encouraged to travel in the other lanes, but are permitted to use the truck lanes. The Tchoupitoulas Corridor improvements at the Port of New Orleans included truck-only lanes to provide efficient access to the port while removing heavy truck traffic from surrounding neighborhoods; the lanes are specifically built to handle the stresses created by heavy truck traffic. In addition, although not a true example of truck-only lanes, a 34-mile segment of the New Jersey Turnpike provides a "dual-dual alignment," in which interior express lanes are reserved for auto-only use and exterior lanes for use by all traffic. A study conducted by the Southern California Association of Governments found that truck only lanes are most feasible under the following conditions:

---

[60] PPP arrangements are increasingly common in rail infrastructure projects because most of the "easy" rail capacity improvements have already been built, leaving only the expensive mega-projects that railroads have difficulty funding through their own cash flow.

– Trucks make up 30 percent or more of the traffic mix;

– Peak-hour traffic volumes are greater than 1,800 vehicles per lane-hour; and

– Off-peak traffic volumes are more than 1,200 vehicles per lane-hour.[61]

- **Short-Sea Shipping** – Short-sea shipping is the movement of goods by water on routes that do not cross an ocean. It is being used in some areas as a strategy to reduce highway congestion by shifting some freight to marine modes via coastal shipping. The United States Maritime Administration (MARAD) recently launched a Marine Highways program to promote increased use of domestic water-borne transportation, including short-sea shipping. The Energy Independence and Security Act of 2007 directed the U.S. Department of Transportation to create a program to expand the use of Marine Highways by designating certain corridors as extensions of the sur-face transportation system and supporting projects that relieve congestion and improve air quality.[62] Short-sea shipping already is used extensively in coastal regions, primarily for bulk commodities like aggregates and fertilizer that are not time-sensitive. One study on the West Coast found that there already was considerable short-sea carrying capacity on vessels making port rotations, but that high terminal and drayage costs would limit the adoption of this mode for coastal shipments.[63] Capital costs to start up short-sea shipping ser-vices also can be a barrier, due in part to Federal requirements (Jones Act) to buy U.S. built vessels for domestic shipping – which can dou-ble or triple the cost of acquiring vessels, as compared to foreign-built vessels.[64] Variations of this strategy include efforts to shift freight to container-on-barge and truck-on-barge modes. Although short-sea shipping has the potential to alleviate pollution and congestion, it can be difficult for it to compete with trucks, particularly for more valu-able, time-sensitive commodities.

---

[61] Southern California Association of Governments and KAKU Associates, *SR 60 Truck Lane Feasibility Study: Final Report,* November 2000.

[62] http://www.marad.dot.gov/ships_shipping_landing_page/mhi_home/mhi_home.htm.

[63] International Mobility Trade Corridor and Cambridge Systematics, Inc. *Cross Border Shortsea Shipping Study,* May 2004.

[64] U.S. Department of Transportation, 2006, *Four Corridor Case Studies of Short-Sea Shipping Services: Short-Sea Shipping Business Case Analysis,* prepared by Global Insight for U.S. DOT Office of the Secretary, August 15, 2006.

3.2.3 OPERATIONAL CHANGES

Freight generators (such as ports and freight-dependent business) and transportation agencies also can change their operating practices in ways that reduce emissions. These strategies can be implemented through regulation, use of new technology, or partnerships with the private sector.

- **Freight Vehicle Speed Reduction Programs** – In order to minimize transit times, freight vehicles typically travel as fast as economically practicable within legal limits. However, vehicles exceeding their most fuel-efficient speed also emit more pollutants per mile traveled. Some jurisdictions have responded by implementing programs to reduce the speed of freight vehicles. These efforts are usually targeted at two modes:

  - **Marine vessels** – Some ports are now requiring vessels to reduce their speed when they come within a certain distance of shore. As in many areas of emissions reduction, California is a leader in the implementation of this strategy. The Ports of Los Angeles and Long Beach have a voluntary speed reduction program in which vessels approaching the port are encouraged to reduce their speed to 12 knots within 20 nautical miles of Point Fermin. Although most ships still operated above 12 knots after the program went into effect, data collected by the Port showed a significant reduction in average ship speeds (from 16 knots to about 13).[65] The California Air Resources Board currently is exploring ways to implement a statewide initiative, either voluntary or regulatory.

  - **Trucks** – Truck speed reduction can be accomplished in a number of ways, including driver training and electronic engine controls. Studies have found that tractor-trailers operating at 55 mph consume up to 20 percent less fuel than trucks driving at 65 mph.[66] This not only reduces emissions, it also saves on fuel and maintenance costs, which can outweigh productivity losses incurred by operating trucks at slower speeds. New truck engines are usually already electronically controlled and trucks can be custom ordered with maximum speed settings built in; existing engines also can be retrofitted with governors (electronic devices that limit maximum speed). Many fleet managers opt to combine this technology with driver training to encourage lower speeds.[67]

---

[65] Garrett, T.L. "Voluntary Commercial Cargo Ship Speed Reduction Emission Reduction Program." Presentation to the CARB Marine Technical Resources Group, December 6, 2001.

[66] U.S. Environmental Protection Agency SmartWay Transport Partnership, "A Glance at Clean Freight Strategies: Reducing Highway Speed," February 2004.

[67] Recent increases in the price of diesel fuel combined with growing corporate environmental awareness has encouraged more trucking firms to adopt these strategies in recent years.

- **Driver Training** – There are numerous driving techniques (in addition to lower speeds) that truckers can employ to reduce emissions and save fuel. Effective trip planning (including alternate routing in the case of construction or emergencies), avoiding rapid acceleration and deceleration, and up shifting as soon as practicable are all ways that drivers can improve fuel economy while reducing freight vehicle emissions. Many trucking firms employ incentive programs that pay drivers bonuses for conserving fuel; engine monitoring systems can be employed to track performance and make recordkeeping easier.

- **Reduced Pickup and Drop-Off Idling for Trucks** – Minimizing time spent idling during pickups and deliveries is another way to reduce emissions, particularly for delivery trucks operating in urban areas where they are likely to make several stops each day. Many freight-generating businesses have adopted no-idle policies at loading facilities in partnership with the EPA through the SmartWay Transport program.[68] Some seaports, including Los Angeles and Long Beach, have implemented gate appointment systems whereby truckers are given a specific time window to pick up a container from the terminal. This strategy reduces unnecessary truck idling at the port gates. Gate appointment systems can be particularly effective since most of the drayage trucks used to move containers short distances (for example, from the port to an intermodal rail yard) are older and more polluting.

- **Off-Peak Cargo Moves** – Another option is to encourage off-peak cargo moves, to reduce congestion and idling both at the port gates and on nearby roadways. The PierPASS/OffPeak program at the Ports of Los Angeles/Long Beach implements such a program by imposing peak-hour fees on cargo handled at the ports (the fees are then used to staff the port during off-peak hours), and providing refunds for loads handled at off-peak hours. However, the effectiveness of off-peak incentives can be limited by customer business hours; many businesses are only open to accept deliveries during the daytime.

- **Improved Port Operational Strategies** – There also are port operational strategies that can be employed inside the terminal gates to reduce truck VMT and emissions. Some container terminals now require trucks to be fitted with radio frequency identification (RFID) tags so that the position of the truck can be monitored within the terminal, enabling terminal operators to better direct truckers to the appropriate place to pick up their cargo. Another strategy is to maintain a chassis pool for truckers who are dropping off or picking up

---

[68] SmartWay is a cooperative program run by EPA that advises companies in the freight sector on how to reduce their emissions and fuel consumption.

containers for more than one shipping line. Most terminals own their own fleet of container chassis, so when a trucker needs to pick up his next load from another terminal he usually must find another chassis owned by the second shipping line. A shared pool of chassis eliminates this problem. The Port of Virginia, for example, contracts with a third party to maintain a chassis pool.

- **Weigh Station Bypass** – Most states weigh trucks operating on major highways to ensure that they are not violating weight restrictions. While this is a necessary function to help prevent excessive pavement and bridge deterioration, it also results in emissions from idling trucks in the weigh station queue. There are two techniques that state DOTs and public safety agencies can adopt to eliminate this problem:

  - **Virtual Weigh Station/Smart Roadside.** Virtual weigh stations, or high-speed weigh-in-motion screening, are remote, unstaffed weigh stations. Typically, weigh-in-motion devices measure and record truck axle weight and gross vehicle weight as the vehicle moves over a sensor installed in the pavement, while a camera is used to identify the vehicle. This information is used to make a screening decision on whether the vehicle should be intercepted (by an enforcement officer) for weighing or inspection. Such systems prevent unnecessary idling at weigh stations and provide continuous data rather than the sample data collected at static weigh stations. They also can minimize scale avoidance, since drivers may not know where the VWS is set up.

  - **Electronic credentialing services** allow trucks equipped with special transponders to bypass weigh stations, port-of-entry facilities, and agricultural inspection stations. For instance, PrePass, one of the most widely available such systems (currently in 29 states), monitors vehicle credentialing and safety and can be used in conjunction with weigh-in-motion devices to ensure compliance with weight requirements.

- **Reduced Empty Mileage** – Empty mileage refers to unloaded truck or rail car movement. This is doubly important to freight carriers, since an empty vehicle incurs costs without earning revenue. Trucking firms and owner-operators can combat this problem in a number of ways, such as hauling loads in a triangular pattern, coordinating with other companies to find backhaul opportunities, purchasing better routing software, and using load-matching sites on the Internet. Implementing flexible shipping and receiving schedules (e.g., 24/7 shipping and receiving) can minimize idling and loading times by avoiding peak hours, but this must be closely coordinated with customers. Rail backhaul is much more difficult because of the inflexible nature of rail routing, but can be accomplished with close coordination between two or more shippers and the railroads serving them.

• **Reduction of Circuitous Train Routing** – Freight trains have very limited routing options because of the fixed nature of their routes and the high capital costs of building new rail corridors. As a result, trains must often take circuitous routes to get between two points, particularly when operating over tracks owned by another railroad through a lease arrangement. The Heartland Corridor Clearance project (described previously) is an example of a capacity project that eliminates a circuitous train route.

• **Construction or Expansion of Truck/Rail Intermodal Facilities** – Truck/rail intermodal transportation combines traditional trucking with line-haul rail service, maximizing the advantages of both modes (lower cost for rail, speed/route flexibility for trucks). In recent years, escalating fuel costs have created more demand for intermodal services, which in turn necessitates the construction or expansion of intermodal facilities to handle the transfer of goods between truck and rail. The Class I railroads have built several new intermodal yards around the country, often with support from local and/or state governments that recognize the economic development and job creation benefits associated with them. For example, BNSF is building a new intermodal facility in Gardner, Kansas, about 25 miles southwest of Kansas City. The yard is the third in a series of 'logistics parks' operated by BNSF in which warehouses and distribution centers are developed adjacent to an intermodal rail yard so that companies can better take advantage of efficient freight rail service. Johnson County and the City of Gardner are making strategic road improvements around the facility, while the Kansas Department of Transportation is improving a key interchange on I-35 that will serve the intermodal terminal. Although much of the activity at the facility will simply be relocated from Kansas City, the new terminal will use electric-powered container handling equipment instead of the diesel-powered vehicles in use at the existing Kansas City yard. In addition, reduced congestion at the new facility is expected to create air quality benefits for the region. If located properly, intermodal yards offer the chance to move goods closer to their final destination via rail or ship (which consume less energy on a ton-mile basis) and then use trucks for the final short-haul trips.

• **Truck Fleet Operational Strategies** – There are a number of strategies that truck fleet owners and manager can employ to reduce transportation expenses which often also reduce fuel consumption and, therefore, emissions. Better alignment of supplier ship points to distribution centers, moving more cargo per trailer, reducing or eliminating unnecessary packaging, and just-in-time logistics all help to minimize distance traveled and/or fuel use. By combining shipments, firms can employ the least-cost transportation option (for example, less-than-truckload shipping is more expensive than truckload, which in turn is more expensive than intermodal). This can be accomplished by combining multiple purchase orders, synchronizing order days, and establishing consistent lead times across company departments shipping from the same location. Knowing the weight and dimensions of items to be shipped helps maximize trailer productivity.

4.0 Funding and Financing Tools for Freight Air Quality Improvements

No transportation project, no matter how beneficial, can proceed without funding. This section describes the major funding and financing options currently available for freight-related air quality improvements. The primary focus is on the suite of Federal tools and strategies available from the U.S. DOT and the U.S. EPA. This section also discusses state and local programs, though in less depth since these vary widely across the country.

4.1 STATE OF THE PRACTICE

There are a number of ways to fund or finance freight air quality projects in the United States, though some are more widely used than others. Broadly speaking, existing funding and financing programs fall into one of four categories:

- **Federal-Aid Highway and Rail Programs.** These are grant programs administered by the U.S. DOT that provide Federal transportation funds for projects that meet the criteria of a given funding program. For freight air quality purposes, the best known of these is the Congestion Mitigation and Air Quality Improvement Program (CMAQ), which also is the only Federal-aid highway program specifically targeted at addressing air quality issues. However, other programs have been successfully used to address freight transportation emissions.

- **Federal Financing Tools.** These are credit facilities that allow sponsors of transportation projects to access capital in order to fund new infrastructure and/or equipment. These programs take the form of loans, credit enhancement, or debt financing and typically feature attractive terms such as low interest rates and long repayment periods. Although no financing tools are targeted directly towards improving air quality, these programs can apply to projects that have air quality benefits, and a few explicitly consider environmental benefits as part of the evaluation process.

- **Other Federal Programs.** Besides the traditional U.S. DOT grant and loan programs, there are some other Federal programs administered by the EPA that provide support for diesel retrofits and other emissions reduction efforts.

- **State, Local, and Nonprofit Programs.** Some state and local resource agencies have their own programs geared specifically towards reducing freight-related diesel emissions. In addition, there are nonprofit advocacy groups and public-private partnerships that focus on air quality. State programs typically take the form of grants or tax credits, while nonprofits sometimes offer loans or credit enhancement.

The following sections provide brief descriptions of these funding and financing tools and the types of freight air quality projects that could be funded with each. For the Federal programs, each section begins with a summary table that briefly describes the key features of each funding program. Readers seeking more information about a particular funding mechanism can then find the detailed description in the text. The state, local, and nonprofit funding opportunities are presented as examples. Rather than providing a comprehensive list of all such programs, the goal is to give readers a sense of the types of programs that may be available.

4.2   FEDERAL-AID HIGHWAY AND RAIL PROGRAMS

Federal-aid highway and rail programs vary widely in terms of scope, purpose, eligibility, and funding levels; as a result, their potential applications to freight air quality improvement also vary widely. An overview of each program is provided below, including a description of the program, project eligibility, and state and local share requirements. In addition, the key advantages and challenges associated with each program are discussed to give practitioners a sense of the practical issues that might be encountered when applying for funds. Table 4.1 summarizes the current U.S. DOT funding Programs.

Table 4.1   U.S. DOT Funding Programs

Funding Program	Eligibility	SAFETEA-LU Funding Level (FY 2005-2009)	Freight Air Quality Application	Project Size	Who Approves Funding?
Surface Transportation Program (STP)	Funds projects on any Federal-aid highway, bridge projects on any public road, transit capital projects, and other state or local projects. Can be used for highway improvements to accommodate rail freight. Project selection criteria vary by state.	$32.6 billion	• Preservation of abandoned rail corridors. • Highway bridge clearance projects to accommodate double-stacked freight trains. • Advanced truck stop electrification systems.	Any size; may require combination with other funding sources for very large projects.	State DOTs/MPOs http://www.transportation.org/?siteid=37&pageid=332 http://www.ampo.org/directory/index.php
CMAQ Congestion Mitigation and Air Quality Improvement Program	Funds transportation projects in nonattainment and maintenance areas that improve air quality. Can be used for start up costs associated with operations (for up to three years).	$8.6 billion	• Advanced truck stop electrification systems • Construction of Intermodal freight facilities that result in air quality improvements on-road and non-road diesel engine retrofits (including Switcher/Shunter Locomotives – acquisition or diesel retrofit and. Drayage Diesel Retrofit. • Cost-effective congestion mitigation activities. Capital investments that enhance air quality, such as on- or near-dock rail, cargo handling equipment, or container on barge facilities • Heavy-duty truck retirement programs. • Railway reconstruction/refurbishment (i.e. for expansion to accommodate transshipped boxes and truckloads)	Any size.	State DOTs/MPOs http://www.transportation.org/?siteid=37&pageid=332 http://www.ampo.org/directory/index.php
Rail Grade Crossings	Provides funding to eliminate rail-highway crossing hazards.	$880 million	• Rail-highway grade separations. • Highway relocation to eliminate crossing. • Rail relocation to eliminate crossing (where most cost-effective).	Small projects; requires combination with other funding sources for very large projects.	State DOTs/MPOs http://www.transportation.org/?siteid=37&pageid=332 http://www.ampo.org/directory/index.php
Truck Parking Facilities Grants	Pilot SAFETEA-LU program; provides grant funds for projects addressing the shortage of long-term parking for commercial vehicles on the NHS.	$25 million	• Construction of commercial vehicle parking facilities adjacent to truck stops and travel plazas. • Constructing turnouts for commercial vehicles. • Improving geometric design of interchanges to improve truck access to parking facilities.	Small project; requires combination with other funding sources for very large projects.	U.S. DOT/FHWA
Capital Grants for Rail Relocation Projects	Provides grants for local rail line relocation and improvement projects. Projects should improve vehicle traffic flow, quality of life, and economic development.	$1.4 billion authorized ($20 million appropriated in FY 2008; $25 million in FY 2009)	• Relocation of a rail line, with ancillary air quality benefits.	Any size, although legislation requires that at least one-half of the funding is used for projects that are $20 million or less.	U.S. DOT/FRA http://www.fra.dot.gov

4.2.1 Congestion Mitigation and Air Quality Improvement Program (CMAQ)

Overview

The CMAQ program funds transportation projects and programs that improve air quality (by reducing transportation-related emissions) in non-attainment and maintenance areas for ozone, carbon monoxide (CO), and particulate matter (PM_{10}, $PM_{2.5}$). Federal funds typically cover 80 percent of a project's cost.[69] Certain activities, including carpool/vanpool projects, priority control systems for emergency vehicles and transit vehicles, and traffic control signalization receive a Federal share of 100 percent. For 2005-2009, CMAQ was authorized for $8.6 billion in funding.

CMAQ funds are apportioned by statutory formula each year to states based on the severity of their ozone and CO pollution. Each state is guaranteed a minimum apportionment of one-half of one percent of the year's total program funding. Individual CMAQ projects are selected by the state or MPOs.

Criteria and Eligibility

While CMAQ is not geared exclusively towards freight projects, many eligible projects are potentially freight-related. Routine maintenance projects and those that expand highway capacity are not eligible for funding, since they do not meet the program's goal of reducing emissions. As with all Federal-aid funding programs, to be eligible, a project must be included in the MPO's current transportation plan and TIP (or STIP for areas not represented by an MPO).

Proposals should include a detailed description of the project, including size, scope, location, and timetable, as well as an air quality analysis that quantifies anticipated emissions reductions that would result from project implementation. If it is difficult to quantify these benefits, a qualitative assessment may be made.

Key Advantages and Challenges

CMAQ funds have been used for a variety of freight-related projects that improve air quality by reducing truck emissions. Examples of CMAQ-funded freight projects include construction of intermodal facilities for moving containers off of highways and onto rail, moving containers

---

[69] Pub. L. 110-140, Sect. 1131, The Energy Independence and Security Act of 2007 provided a 100 percent Federal share for CMAQ projects in FY 2008 and 2009. Continuation of this provision with new transportation legislation is unknown.

off highways by defraying barge operating costs, rail track rehabilitation, diesel engine retrofits, idle-reduction projects, and new rail sidings. Additionally, though previously eligible, SAFETEA-LU highlighted advanced truck stop electrification systems (on non-Interstate right-of-way),[70] on-road diesel engine retrofits, heavy duty truck retirement programs, and other cost-effective congestion mitigation activities as CMAQ eligible projects. SAFETEA-LU also established eligibility for non-road diesel engine retrofit projects (container gantry cranes, switcher/shunter locomotives). SAFETEA-LU directs states and MPOs to give priority to diesel retrofits and cost-effective congestion mitigation activities.

CMAQ may be used to fund construction and other activities that could benefit a private entity, if it can be documented that the project will remove truck traffic on the Federal-aid system or reduce other freight-related emissions, thus improving the region's air quality. This would be accomplished through a public-private partnership (PPP) agreement, the mechanism that allows spending public CMAQ funds on most private freight projects. Agencies considering the use of a PPP approach should develop a contract or memorandum of understanding that ensures a public air quality benefit is achieved.

The primary advantage of CMAQ for freight and air quality is that it is specifically targeted at projects that improve air quality. However, it is not freight-specific, so freight air quality projects must often compete for limited funds with other non-freight improvements. This is compounded by the fact that many states and MPOs have not fully integrated freight into their transportation planning and programming activities, so freight sometimes does not have a strong "voice" in project evaluation and selection.

The multijurisdictional nature of freight movements also poses a challenge for obtaining freight project funding under CMAQ. CMAQ funds are usually devoted to projects that have benefits in a specific nonattainment area, but freight nearly always moves in and out of a given metro area, which can interfere with CMAQ eligibility. Certain types of freight projects – such as intermodal freight terminals – remain eligible despite this fact. Others (like APUs on trucks) may not be given priority unless the vehicles are expected to remain in the nonattainment area the majority of the time. (In some cases, project sponsors require equipment, such as railyard switching engines, to stay in the nonattainment area for a certain period of time as a condition of receiving CMAQ funds.) Project sponsors should, therefore, make certain that a given project is eligible prior

---

[70] A provision for construction of truck stop electrification facilities in the Interstate ROW [U.S. Code Title 23, Section 111(d)] was removed in the SAFETEA-LU Technical Corrections Bill.

to developing an application for funding. In any event, sponsors need to show that sufficient air quality benefits will be realized within the nonattainment area in order to qualify for CMAQ funds.

4.2.2   SURFACE TRANSPORTATION PROGRAM (STP)

Overview

The STP provides a flexible source of funding that states can use for projects on any Federal-aid highway, bridge projects on any public road, transit capital projects, and intercity/intracity bus terminals and facilities, among other things. Funds are apportioned to states based on three primary factors:

- Total lane-miles of Federal-aid highways (25 percent);

- Vehicle-miles traveled on lanes on Federal-aid highways (40 percent); and

- Estimated tax payments contributed by state highway users to the Highway Account of the Highway Trust Fund (35 percent).

Notwithstanding the above criteria, each state receives a minimum of one-half of one percent of the total funds apportioned for the STP.

These funds may be used for a wide variety of transportation investments, including certain types of preventive maintenance[71] new road construction, and transit capital projects. SAFETEA-LU expanded eligibility to include advanced truck stop electrification projects. The Federal share for projects funded under STP is generally 80 percent.

Criteria

STP project selection criteria vary among states and MPOs, since transportation agencies can develop their own evaluation criteria subject to the planning requirements set out in statute.[72] As a result, project sponsors need to consider STP selection criteria in their state or region in light of how those criteria would (or would not) apply to a freight air quality project. Not all states or MPOs specifically plan for freight, and as a result freight projects can have a hard time competing for scarce transportation funding.

---

[71] STP funds are only eligible for certain types of preventive maintenance and are never eligible for routine maintenance. See http://www.fhwa.dot.gov/preservation/100804.cfm.

[72] 23 CFR Parts 450 and 500, 49 CFR Part 613.

Key Advantages and Challenges

Freight projects eligible for STP money include:

- Preservation of abandoned rail corridors;

- Highway bridge clearance increases to accommodate double-stack freight trains; and

- Capital costs of advanced truck stop electrification systems not in an Interstate right-of-way but on an eligible publicly owned location.

Although truck stop electrification is the only category that is specifically targeted toward emissions reduction, the other project types often have ancillary air quality benefits, particularly if they create a mode shift from truck to rail. In this way, air quality benefits may be used to help move a freight project forward.

Although STP funds may be used for pollution abatement projects, that is just one of many potential uses for these funds. In addition, STP funds are suballocated in various ways, with certain portions going towards urban and rural areas, transportation enhancement, and other subcategories. As a result, air quality projects are not often considered for funding under STP, since most states prefer to use the money for new capacity, or other activities such as sidewalk construction and to steer air quality projects to the CMAQ program. Combined with the fact that a lot of states and MPOs still do not actively push freight projects from planning stages to implementation, this can make freight air quality projects a long shot in many places.

Because of the wide eligibility of STP projects, STP funds tend to get used up more quickly than other sources such as CMAQ. In addition, anti-idling efforts are essentially the only air quality technology strategy allowed under STP. It also is important to note that other idle-reduction strategies (such as shore power systems and idle-limiters) are not eligible under STP unless they are identified as Transportation Control Measures pursuant to Section 108 of the Clean Air Act. Project sponsors should make certain that a given project is eligible for STP funds prior to developing an application for funding.

Finally, some states restrict the use of STP for non-highway projects, which may make rail, marine, and air quality improvements ineligible.

4.2.3 NATIONAL HIGHWAY SYSTEM (NHS)

Overview

The NHS is composed of certain roadways identified as being critical to the nation's economy, defense, and mobility. The system currently includes about 160,000 miles of roads throughout the country. It is defined in statute[73] and includes the Interstate system, the Strategic Highway Network (StraHNet), other Principal Arterial roadways not designated as part of the Interstate or StraHNet systems, highway connections between major military facilities and StraHNet, and designated intermodal connectors. NHS funds are distributed to states by formula allocation; SAFETEA-LU authorized $30.5 billion for the NHS for FY 2005 to 2009.

Generally, NHS projects receive 80 percent Federal funding with the remaining 20 percent coming from state and/or local sources. (Certain activities, such as HOT lanes and safety projects, receive a higher Federal share.)

Criteria and Eligibility

Eligible freight projects include construction, reconstruction, resurfacing, and rehabilitation on an intermodal connector – a designated roadway connecting the NHS with a truck-rail facility, port, pipeline terminal, or an airport. So, for example, a state could use NHS funding to improve a truck access route to a port, which could remove a bottleneck and thus reduce emissions. However, only designated NHS intermodal connectors are eligible for funding. The lists of these corridors are only updated intermittently since designation is largely a state responsibility. In addition, NHS connectors account for only a small percentage of the total number of connectors (many connectors do not meet the eligibility requirements for the NHS).[74] This means that the freight facilities that qualify for funding under this program are quite limited.

Key Advantages and Challenges

Intermodal access projects with air quality benefits (such as ones that improve traffic flow on an intermodal connector) could make use of NHS funding. Of course, freight projects seeking NHS funding must compete with non-freight projects, much like those applying for STP funds.

---

[73] 23 U.S.C. 103(b)(2).

[74] NHS Intermodal Connector Requirements may be found at http://ops.fhwa. dot.gov/freight/freight_analysis/nhs_intermod_fr_con/app_a.htm.

4.2.4 TRUCK PARKING FACILITIES GRANTS

Overview

Truck Parking Facilities Grants is a pilot program that provides grants for projects that address the shortage of long-term parking for commercial vehicles on the NHS. SAFETEA-LU authorized $25 million over five years for the program, but after take-down charges and Congressional rescissions approximately $17 million was appropriated through fiscal year 2009. Fiscal Year 2010 provided another $5.84 million. Eligible projects include construction of new or expanded commercial vehicle parking facilities, construction of turnouts for commercial vehicles, improvement to interchanges, and Intelligent Transportation System (ITS) deployments promoting availability of parking.

Criteria and Eligibility

Applicants must describe the safety benefits that will result from the proposed project as well as mobility improvements and congestion relief. Applications are scored based on the following criteria:

- Demonstration of severe shortage of commercial motor vehicle parking capacity/utilization in corridor or area to be addressed;

- Extent to which the proposed solution resolves the shortage;

- Cost-effectiveness of the proposal; and

- Scope of the proposal, including evidence of input from a wide range of affected parties such as community groups, local governments, MPOs, and motorist/trucking organizations.

According to the program language, funding priority is given to applicants that "demonstrate that their proposed projects are likely to have positive effects on highway safety, traffic congestion, or air quality."[75] However, air quality is not one of the evaluation criteria.

4.2.5 RAIL-HIGHWAY CROSSING PROGRAM

Overview

---

[75] FHWA Truck Parking Facilities Fact Sheet.

This program was previously a set-aside of the Surface Transportation Program (STP). It provides funding for projects that improve safety at railroad crossings by eliminating hazards (e.g., grade separation, or highway or rail relocation to eliminate a crossing) and/or installing or upgrading crossing devices. SAFETEA-LU funding from FY 2006-2009 for this program is $880 million.

Each state receives a minimum of one-half of one percent of program funds. One-half of the apportioned funds are distributed based on formula factors for the Surface Transportation Program; the other one-half are distributed according to the number of public railway-highway crossings in each state. The legislation requires states to set aside at least one-half of their funding allocation for the installation of protective devices at rail-highway crossings. If all needs for installation of protective devices have been met, then the funds available can be used for other at-grade crossing projects eligible under this program. Only safety improvements are eligible. The Federal share of project funding is 90 percent.[76] Projects are approved for funding by their respective state DOTs and/or MPOs.

Criteria and Eligibility

Eligible projects include:

• Separation or protection of grades at crossings;

• The reconstruction of existing railroad grade crossing structures; and

• The relocation of highways or rail lines to eliminate grade crossings.

States are responsible for determining which public crossings need improvements. Specific safety improvements could also reap air quality benefits, although these benefits are not themselves eligible for this funding. Project sponsors could, therefore, evaluate air quality benefits among the set of safety improvements being considered to determine if there are additional benefits worthy of consideration. It is important to note that SAFETEA-LU requires that states set aside at least 50 percent of the funding allocation for the installation of protective devices at rail-highway crossings. Only after all needs for installation of protective devices have been met can funds be used for other at-grade crossing projects eligible under this program.

---

[76] The Federal share may be 100 percent for certain improvements such as crossing closures, hazard elimination, signing, pavement markings, and active warning devices.

Key Advantages and Challenges

This program could be used to fund grade separations that may have an ancillary air quality benefit. These emission reductions are generally achieved by reduced idling and congestion on the roadways that cross the tracks. In the vicinity of a major intermodal facility, much of this traffic also may be freight trucks.

Theoretically, a state could use the emissions reduction expected from a grade separation project as a factor in project evaluation and selection. However, the Rail-Highway Crossing Program is narrowly focused on safety, and only safety improvements are eligible. Therefore, a grade separation project would have to be justified primarily on safety grounds to obtain funding under this program.

4.2.6 CAPITAL GRANTS FOR RAIL LINE RELOCATION

Overview

This program, established under SAFETEA-LU and administered by the FRA, provides grant funding for local rail line relocation and improvement projects that improve rail safety, motor vehicle traffic flow, community quality of life, or economic development, or involve the relocation of any part of the rail line. Under the legislation, $1.4 billion was authorized for these projects ($350 million per fiscal year), subject to appropriations. However, Congress did not appropriate any money for this program for FY 2006 or 2007. In FY 2008, Congress appropriated approximately $20 million for this program, $5.25 million of which was earmarked for nine noncompetitive projects. For FY 2009, $25 million were appropriated by Congress, with $17.1 million directed to 23 projects. Projects can be of any size, but the legislation requires that at least one-half of the total funding be used for projects costing $20 million or less. The Federal share of funding for any given project cannot be more than 90 percent. Only construction costs are reimbursable under this program. This includes architectural and engineering costs.

Criteria and Eligibility

The FRA issued a Final Rule outlining the regulations governing this program in July 2008.[77] According to the Final Rule, the criteria to be considered when selecting projects for funding include:

---

[77] "Implementation of Program for Capital Grants for Rail Line Relocation and Improvement Projects," Federal Register Volume 73, No. 134, Friday, July 11, 2008.

- The ability of the state requesting the grant to fund the project without Federal assistance, as measured by:

    - The existence of state programs to improve railroads;

    - The state's use of available highway-rail grade crossing improvement funds provided through 23 U.S.C. 130; and

    - Other indicators of creditworthiness, such as bond ratings.

- The allocation requirements mentioned above;

- Equitable treatment of the various regions of the country;

- The effects of the project on motor vehicle or pedestrian traffic, safety, community quality of life, and area commerce;

- The effects of the project on freight and passenger rail operations on the rail line; and

- Any other factors the FRA deems to be relevant in assessing the effectiveness of the proposed improvement in achieving the goals of the national program. This may include any air quality benefits associated with a rail line relocation. States are required to submit a complete description of the expected public and private benefits associated with their projects.

Key Advantages and Challenges

Like the Rail-Highway Crossing Program, Capital Grants for Rail Line Relocation theoretically could be used for an air quality project. Emissions reductions might score highly on the community quality of life criteria and presumably would be included in the applicant's list of public and private benefits. However, there are many other factors that the FRA considers when evaluating applications, including safety, emergency vehicle access, traffic counts at highway crossings, and the effects of the relocation on local industry (both positive and negative). To the extent that these priorities would compete with any air quality benefits, this could limit the applicability of this program for emissions reduction efforts.

4.3 FEDERAL FINANCING TOOLS

Beyond traditional Federal-aid grant programs, there are a number of Federal financing tools available for freight-related air quality improvements. Many are targeted towards specific modes (such as rail) that sometimes get overlooked under grant funding programs. Below are detailed descriptions of the financing tools, their requirements, and eligible projects. Table 4.2 presents summary information about them.

Table 4.2   Federal Financing Tools

Funding Program	Eligibility	SAFETEA-LU Funding Level (FY 2005-2009)	Freight Air Quality Application	Project Size	Who Approves Funding?
Transportation Infrastructure Finance and Innovation Act (TIFIA)	Provides loans and credit assistance for major transportation investments of national or regional significance, including public intermodal freight facilities. SAFETEA-LU expanded TIFIA eligibility to private rail projects. Private sponsors are eligible.	SAFETEA-LU authorizes $122 million per year to pay the subsidy costs of supporting Federal credit under TIFIA. This level of funding can support loans with a total value of more than $2 billion annually.	• Public or private rail facilities providing benefits to highway users • Intermodal freight transfer facilities • Access to freight facilities and service improvements, including ITS • Surface transportation infrastructure modifications to facilitate intermodal interchange, transfer, and access into and out of ports	$50 million minimum, no specific maximum; air quality benefits would be ancillary to capacity expansion project	U.S. DOT http://tifia.fhwa.dot.gov
State Infrastructure Banks (SIB)	SAFETEA-LU authorizes all 50 states, the District of Columbia, Puerto Rico, and U.S. territories to establish infrastructure revolving funds that can be capitalized with Federal transportation funds authorized through FY 2009. Current legislation allows for the creation of rail accounts. Private sponsors are eligible.	Highway Account – Up to 10 percent of NHS, STP, Bridge, and Equity Bonus programs, at the discretion of the state DOT. Rail Account – Funds made available for capital projects under Subtitle V (Rail Programs) of Title 49.	• Truck stop electrification • Certain rail capital projects • Other applications as determined by state enabling legislation	Any size; depends on state capitalization. Generally small projects are funded.	State DOT (and/or SIB Board established). http://www.transportation. org/?siteid=37&pageid=332
Railroad Rehabilitation and Improvement Financing (RRIF)	Loans and credit assistance to both public and private sponsors of rail and intermodal projects. Private sponsors are eligible.	$35 billion; $7 billion is directed to shortline and regional railroads.	• Locomotive rehab or purchase to improve fuel economy/productivity and reduce emissions • "Green" locomotive purchase	Generally small projects; emissions reductions are usually an ancillary benefit.	U.S. DOT/FRA http://www.fra.dot.gov
Private Activity Bonds	Title XI Section 1143 of SAFETEA-LU amends Section 142(a) of the IRS code to allow the issuance of tax-exempt private activity bonds for highway and freight transfer facilities. Private sponsors are eligible.	Up to $15 billion.	• Rail-truck transfer facilities • Port access projects • Air quality initiative as part of a larger infrastructure expansion	Any size; potential for large infrastructure projects.	U.S. DOT http://www.fhwa.dot.gov/ ipd/p3/tools_programs/ pabs.htm
GARVEE Bonds	Financing instrument that allows state to issue debt backed by future Federal-aid highway revenues. Eligibility for freight projects is constrained by the underlying Federal-aid programs that will be used for debt service.	N/A	• Any application that is authorized by the underlying grant programs	Typically large projects or groups of projects ($10 million or larger).	State DOT/Local Government must be willing to dedicate future revenue. http://www.transportation. org/?siteid=37&pageid=332

Table 4.2   Federal Financing Tools (continued)

Funding Program	Eligibility	SAFETEA-LU Funding Level (FY 2005-2009)	Freight Air Quality Application	Project Size	Who Approves Funding?
SEP-15	Experimental process for FHWA to evaluate new approaches to project delivery involving public private partnerships. The Secretary of Transportation is authorized to waive certain regulations on a case-by-case basis. PPP arrangements may involve innovative financing.	N/A	• Air quality benefits likely to be ancillary to a capacity improvement; planners should note additional benefits that would accrue if project were accelerated	Typically large capacity expansions.	http://www.fhwa.dot.gov/ ipd/p3/tools_programs/ sep15.htm
Section 129 Loans	Provides Federal funding to public or private project sponsors for tolled or free roads that are not part of the Interstate system. Loans extended by states are treated as eligible Federal-aid project costs; when they are repaid, states can use the funds on other Title 23 eligible projects.	N/A	• Truck access route to a port or intermodal terminal • Only for highways, bridges, and tunnels, but repaid loan amounts may be used for any Title 23 eligible project	No limit, but loans are limited to 80 percent of eligible project costs.	State DOT http://www.fhwa.dot.gov/ ipd/finance/tools_programs/ federal_credit_assistance/ section_129/index.htm

4.3.1 RAILROAD REHABILITATION AND IMPROVEMENT FINANCING (RRIF)

Overview

This program provides loans and loan guarantees to public and private sponsors of rail and intermodal transportation investments. The funds may be used to acquire, improve, or rehabilitate intermodal or rail equipment and facilities, or to refinance debt incurred for these activities. They also can be used to develop new intermodal or railroad facilities. RRIF loans may not be used for operating expenses.

SAFETEA-LU expanded the RRIF program tenfold, from $3.5 billion to $35 billion, with $7 billion reserved for shortline and regional railroads. The legislation also specifically added rail infrastructure and rail bottleneck relief to the list of program priorities.

Loans may be used to fund up to 100 percent of a project; repayment periods are up to 25 years with interest rates equal to the cost of borrowing to the government. The program is managed by the Federal Railroad Administration (FRA).

Criteria and Eligibility

RRIF applications are scored based on the following criteria:

- **Eligibility of the Applicant** – Eligible applicants include railroads, state and local governments, government-sponsored authorities and corporations, joint ventures that include at least one railroad, and limited option freight shippers who intend to construct a new rail connection;

- **Eligibility of the Project** – Priority is given to projects that enable U.S. companies to be more competitive in international markets, are endorsed by existing state rail plans, or preserve or enhance rail or intermodal service to small communities or rural areas;

- **Creditworthiness of the Applicant** – Creditworthiness is measured by financial statements, financial projections, and a credit rating from a nationally recognized rating agency;

- **Extent to which the Project will Enhance Safety** – How the project will contribute to safe railroad operations for both rail employees and the public;

- **Significance of the Project** – In terms of generating economic benefits and improving the rail transportation system;

- **Improvement to the Environment** – Any environmental benefits that are expected to result from the project, including air quality benefits/ emissions reductions; and

- **Improvement in Service or Capacity** – Any anticipated service improvements in the railroad system, or the reduction of service or capacity problems expected to result from the project.

Key Advantages and Challenges

As noted above, environmental benefits are considered when the FRA evaluates applications for RRIF loans, and the program can be (and has been) used to finance rail equipment purchases that reduce emissions. Furthermore, RRIF loans are not subject to the geographical constraints that can make it difficult to fund similar projects with CMAQ dollars.

As a practical matter, however, the assets purchased must generate a positive financial return that would allow the railroad to pay the loan back. Fuel savings and increased productivity of cleaner, more efficient equipment may well provide such a return, but the rate of return also is affected by the higher initial costs of the equipment purchase. Since railroads are mostly private, for-profit companies, projects also would be subject to the vagaries of the economic cycle. Railroads would be unlikely to take out a RRIF loan to purchase additional rolling stock (green or otherwise) in a slack economic environment, especially if they already have significant idle capacity.

Finally, it is important to remember that emissions reductions achieved with a RRIF loan are likely to be an ancillary benefit of a larger project. As a result, project sponsors need to understand the conditions under which an additional investment in fuel-saving or idle reduction technology would be attractive to the railroad. This would include factors such as interest rate, repayment term, and potential productivity improvements. Officials should work with their railroad partners to identify these conditions and tailor projects so that they conform to the railroads' business interests while achieving a public air quality benefit.

Two recent examples of RRIF loans used to finance locomotive purchases or rehabilitation illustrate some of these points. In 2008, a regional railroad located in the Midwest received a $31 million RRIF loan. The funds were used to purchase 12 new locomotives, which allowed the railroad to increase train lengths, tonnage, and operating speeds while providing enhanced service to recently constructed ethanol plants. The new locomotives also are more fuel efficient with lower emissions. Similarly, in 2007 a railroad service company with shortline operations in the Midwest and Southeast received a RRIF loan for $59 million, part of which was used to rehabilitate 24 locomotives to increase fuel efficiency, reduce diesel emissions, and improve reliability. In both cases, the railroads included the expected fuel savings and emissions reductions that would result from these projects in their application.

4.3.2 TRANSPORTATION INFRASTRUCTURE FINANCE AND INNOVATION ACT (TIFIA)

Overview

The TIFIA credit program is designed to leverage limited Federal resources and stimulate private capital investment by providing credit assistance (up to 33 percent of the project cost) for major transportation investments of national or regional significance. Credit assistance is provided through secured loans, loan guarantees, or lines of credit.

SAFETEA-LU authorizes $122 million per year to pay the subsidy costs of supporting Federal credit under TIFIA. There is no limit on the amount of credit assistance that can be provided to borrowers in a given fiscal year. Repayment of TIFIA loans is required to come from dedicated revenue sources, such as tolls or user fees. As of April 2009, TIFIA assistance amounted to $6.6 billion, leveraging $24.4 billion in transportation investments for a total of 18 projects. About $994 million in TIFIA debt has been repaid to date. Additional information on this financing program is available at http://tifia.fhwa.dot.gov/.

Criteria and Eligibility

Project costs must be at least $50 million or one-third of the state's annual apportionment of Federal-aid highway funds, whichever is less.[78]

Eligibility for freight facilities include:

- Public or private freight rail facilities providing benefits to highway users;

- Intermodal freight transfer facilities;

- Access to freight facilities and service improvements, including capital investments for ITS; and

---

[78] There is an exception for ITS projects, which must be at least $15 million.

- Port terminals, only when related to surface transportation infrastructure modifications to facilitate intermodal interchange, transfer, and access into and out of the port.

Applications are evaluated based on a set of eight criteria that are specified in statute,[79] each with a specific weight assigned by the U.S. DOT:

- **Significance to the regional or national transportation system** (20 percent), defined as the national or regional significance of the project in terms of generating economic benefits, supporting international commerce, or otherwise enhancing the national transportation system.

- **Private participation** (20 percent), or the extent to which the project fosters innovating public-private financing arrangements that attract private debt or equity investment.

- **Environmental benefits** (20 percent), or the extent to which the project helps to maintain or protect the environment. This includes reductions in air, water, or noise pollution that would not otherwise occur if the project were not built, as well as any major mitigation efforts and whether those efforts go above and beyond what is required by law.

- **Project acceleration** (12.5 percent), which is the likelihood that TIFIA assistance would allow the project to proceed earlier than it would otherwise be able to.

- **Creditworthiness of the project** (12.5 percent), including a determination by the Secretary of Transportation that any financing for the project has appropriate security features, such as a rate covenant, to secure repayment of the loan.

- **Use of technology** (five percent), which is the proposed use of new technologies, such as intelligent transportation systems (ITS), that enhance the project's efficiency.

- **Consumption of budget authority** (five percent), or the amount of budget authority consumed by the project sponsor in funding the requested credit instrument.

- **Reduced Federal grant assistance** (five percent), which is the extent to which credit assistance would reduce the amount of Federal grant assistance required for the project.

---

[79] 23 U.S.C. §602(b)(2).

Because TIFIA focuses on very large-scale capacity projects, applicants must have circulated a Draft Environmental Impact Statement (DEIS) at the time of application for TIFIA credit assistance, unless the project has received a Finding of No Significant Impact or a Categorical Exclusion.[80] The U.S. DOT will not obligate funds under TIFIA before a Record of Decision has been issued for the project following EPA regulations.

Key Advantages and Challenges

Environmental benefits are specifically considered in TIFIA project evaluation, and they receive more weight than most of the other criteria. However, the program is focused on capacity enhancement, particularly large-scale projects that are difficult to fund through traditional means. Although such improvements may well have air quality benefits, they are unlikely to move forward on those grounds alone. Practitioners should, therefore, look for ways to strengthen a potential TIFIA application through the inclusion of an emissions-reduction strategy with quantifiable results. An example might be a port expansion that includes improvements to highway and/or rail access.

4.3.3 STATE INFRASTRUCTURE BANKS (SIB)

Overview

The SIB program allows states to establish revolving funds to pay for infrastructure investments. These funds are capitalized with Federal transportation funds authorized through FY 2009. It is possible to create multistate SIBs, which may be used to finance regional freight improvements that cross state boundaries. Through a SIB, states can lend money to public and private sponsors of transportation projects. SIB funds also may be used to provide credit assistance for projects being financed by other means. When loans are repaid, the funds are "recycled" to finance future transportation investments. Some states (such as Florida) have established state-funded SIBs, which can be integrated with the Federal SIB program.

States participating in the SIB program may capitalize the account(s) in their SIBs with Federal surface transportation funds[81] for each of FY 2005-2009 as follows:

---

[80] A Finding of No Significant Impact (FONSI) means that a preliminary environmental assessment has found that the proposed project would have no significant environmental impacts. Categorical Exclusions are certain types of projects that are exempt from detailed environmental analysis because they are known to have no significant environmental impacts. Most states have lists of project types that are categorical exclusions.

[81] However, a SIB cannot be capitalized with CMAQ funds, although SIB funds can be used for a CMAQ-eligible project.

- **Highway Account** – Up to 10 percent of the funds apportioned to the state for the NHS, STP, Bridge, and Equity Bonus;

- **Transit Account** – Up to 10 percent of funds made available for capital projects under Urbanized Area Formula Grants, Capital Investment Grants, and Formula Grants for Other Than Urbanized Areas;

- **Rail Account** – Funds made available for capital projects under Subtitle V (Rail Programs) of 49 USC; and

- The state must match Federal funds used to capitalize the SIB on an 80-20 Federal/non-Federal basis.

Currently, 31 states and Puerto Rico have SIBs. These states have issued more than $5 billion in loans.

Criteria and Eligibility

Selection criteria vary by state, since it is left up to individual states to develop enabling legislation. All projects must be eligible for Federal-aid under Title 23 or Title 49, United States Code. Projects that are eligible for CMAQ funds also are eligible for SIB funding.

Key Advantages and Challenges

Because they are focused on capacity improvements, SIBs rarely fund projects that are solely focused on freight air quality. However, it is possible to do so. For example, New York State used its SIB to implement two truck stop electrification projects on the New York State Thruway.

A SIB can make loans that many private lenders would consider to be too risky, thus allowing a project sponsor to achieve a public benefit that may otherwise not be realized. A SIB cannot be capitalized with CMAQ funds, but SIB funds can be used to finance CMAQ projects that have a consistent revenue stream which can be used to repay the loan.

SIB structures vary widely by state, so project sponsors should consult their state's enabling legislation to determine which (if any) freight air quality projects can be funded through a SIB. This also would allow practitioners to identify other projects where a SIB application might be strengthened with the addition of an air quality component.

4.3.4 GRANT ANTICIPATION REVENUE VEHICLES (GARVEE BONDS)

Overview

GARVEE bonds are debt instruments issued by a state to finance a transportation investment and backed by the state's anticipated future Federal-aid grant receipts. Under the legislation, a state may be reimbursed for debt service and/or issuance costs with dollars from the state's future Federal-aid highway apportionments.[82] To comply with Federal requirements for a fiscally constrained planning process, the Federal share of debt-related costs anticipated to be reimbursed over the life of the bonds must be designated as Advance Construction (AC) and included in the State Transportation Improvement Program (STIP). In this way, the state's future revenue from Federal-aid grants is converted to immediate funding for needed transportation improvements. FHWA approves only the project to be financed with a bond issue, not the actual issuance of debt, which is under the authority of the state. A SIB can issue GARVEE bonds on behalf of a project. Since its creation, 22 states plus Puerto Rico and the Virgin Islands have issued more than $9.6 billion in GARVEE bonds.

Criteria and Eligibility

Eligibility for freight air quality projects is constrained by the underlying Federal-aid highway programs that will be used to repay debt service. In other words, the project must meet the eligibility criteria for whatever grant program(s) are being used to back up the bonds. Other characteristics of the underlying grant programs, such as matching requirements, also carry over. The AC amount designated when the project is approved must consist of some combination of eligible funding categories. The state does, however, retain the right to decide each year which funding categories to obligate for AC conversion.

Key Advantages and Challenges

There are few (if any) examples of freight air quality projects funded through GARVEE debt issuance. However, due to the nature of the program, any air quality improvement that is eligible for a given Federal grant apportionment also would be eligible for GARVEE funding. So, for example, a state could leverage future CMAQ dollars to make an eligible freight air quality improvement.

---

[82] 23 U.S.C. §122, "Reimbursements to States for Bond and Other Debt Instrument Financing Costs."

GARVEE bonds are one way to leverage future Federal transportation funding to accelerate a freight air quality improvement. They also offer flexibility of funding sources, since as long as a project conforms to the requirements of the underlying Federal-aid programs, it can be funded with GARVEE bonds. However, in the past some states have been cautious about pledging future Federal transportation dollars to fund current projects because they are fearful that they may over commit themselves. Future Federal-aid revenue streams are not guaranteed, and many states have balanced budget requirements that limit their ability to deal with potential over commitments. Project planners should, therefore, find out their state's policy before attempting to use GARVEE bonds for a freight air quality project.

4.3.5 Private Activity Bonds

Overview

States and local governments are allowed to issue tax-exempt bonds to finance highway and freight transfer facility projects sponsored by the private sector, under an amendment to the IRS Code made in Title XI Section 11143 of SAFETEA-LU. SAFETEA-LU includes a cap of $15 billion on private activity bonds.

Passage of the private activity bond legislation reflects the Federal Government's desire to increase private sector investment in U.S. transportation infrastructure. Providing private developers and operators with access to tax-exempt interest rates lowers the cost of capital significantly, enhancing investment prospects. Increasing the involvement of private investors in highway and freight projects generates new sources of money, ideas, and efficiency. Private activity bonds also may be combined with TIFIA credit assistance.

Criteria and Eligibility

There are no specified criteria or eligibility requirements for this program, except that the bond issuer must be a government agency acting as a conduit for the private entity. Because the intent of the program is to attract private capital to transportation improvements, any project using Private Activity Bonds would need to provide a positive financial return sufficient to repay the bonds.

Proposed application requirements issued by the U.S. DOT encourage applicants to submit applications to the DOT with basic information such as project description, amount of bonding authority requested, borrower information, project schedule, financials, and other key information.[83]

---

[83] http://frwebgate.access.gpo.gov/cgi-bin/getdoc.cgi?dbname=2006_ register&docid=fr05ja06-64.

Key Advantages and Challenges

Current allocations total about $6.3 billion for seven projects, but bonds have been issued for only two, neither of which are specifically geared towards freight. However, there are two freight transfer projects that were approved for PAB issuance: The RidgePort Logistics Center in Will County, Illinois and the CenterPoint Intermodal Center in Joliet, Illinois. Applications also have been received for intermodal facilities in Kansas City and Seneca, Illinois. To the extent that these projects would shift freight from trucks to rail, they do offer a substantial air quality benefit.

A project undertaken with private activity bond funding would have to generate a sufficient return on investment to enable the borrower to repay the bond. This might exclude many air quality projects, unless air quality benefits are definitively linked to fuel savings or other economic benefits that would accrue to private sector partners. As with RRIF loans, the terms would have to be attractive enough to the private sector borrower. In the case of "green" technology, repowering, and the like, this would be affected by any extra up-front costs associated with the cleaner equipment.

Given the typical size of PAB-funded projects (each of the approved projects is worth several hundred million dollars), the best approach for a freight air quality initiative may be to integrate it into a larger capacity enhancement, such as a major port expansion.

4.3.6 SECTION 129 LOANS

Overview

Section 129 loans, which are named after the section of United States Code in which they are described,[84] allow Federal participation in loans made by states to toll projects or to non-toll projects that have a dedicated revenue stream. States may extend loans to public or private sponsors of transportation projects; recipients are selected according to each state's specific laws and processes. The loan amount is treated as an eligible Federal-aid project cost, so states can seek reimbursement from the Federal government up to the maximum Federal share (80 percent of eligible project costs). States have broad authority to negotiate loan terms with prospective borrowers.

As with most other loan programs, a dedicated repayment source must be identified and a pledge for repayment secured in advance. States can use repaid loan amounts to fund other Title 23 eligible transportation projects, thus allowing them to "recycle" Federal-aid highway funds. They also can use repayments for additional credit enhancement activities.

---

[84] 23 U.S.C. 129(a).

Loans can be for any amount up to the maximum Federal share. The loan may be used for any phase of a project, with the stipulation that costs incurred prior to loan authorization may not be paid using loan proceeds. Recipients must begin to pay back the loans 5 years after project completion, and repayment must be completed within 30 years of loan authorization.

Criteria and Eligibility

Section 129 loans can be made to any public or private project sponsor that is building or planning to build a Federal-aid eligible toll project or a non-toll highway project with an identified revenue stream for loan repayment. Eligibility requirements are determined by the pot of money the Federal funding originates in – for instance, a Section 129 loan financed with CMAQ funds must go to a CMAQ eligible project. Other specific selection criteria within these eligibility requirements are determined by individual states.

Key Advantages and Challenges

Section 129 loans are designed to provide easily accessible start-up financing for toll roads or other privately sponsored projects. Since repaid amounts can be used on any Title 23-eligible transportation project, states could conceivably attempt to target these "recycled" funds towards freight air quality projects, including CMAQ projects (such as locomotive rehabilitation or replacement), truck stop electrification, and intermodal access projects.

4.3.7 SPECIAL EXPERIMENTAL PROJECT 15 (SEP-15)

Overview

SEP-15 is an experimental process for FHWA to identify, for trial evaluation, new public-private partnership approaches to project delivery. The goal is to allow transportation projects to be delivered quickly while simultaneously protecting taxpayer money and the environment.

Criteria and Eligibility

There are no specific SEP-15 eligibility requirements or selection criteria since the program is intended to develop financing approaches outside of the traditional Federal funding universe. Under the legislation enabling SEP-15, the Secretary of Transportation may waive Title 23 requirements and regulations on a case-by-case basis. SEP-15 allows FHWA to experiment within four major components of project delivery:

- Contracting;

- Compliance with FHWA's internal NEPA process and other environmental regulations;

- Right-of-way acquisition; and

- Project finance.[85]

This experimentation may involve other elements of the transportation planning process and/or deviations from current requirements under Title 23 U.S.C. governing transportation projects or applicable FHWA regulations. Project sponsors under SEP-15 may suggest modifications to traditional FHWA project approval processes.

Key Advantages and Challenges

To date, there have been 12 projects located throughout the country that have been approved for various waivers of requirements or experimental delivery approaches through the SEP-15 program. None are specifically targeted towards air quality improvement; rather, most are capacity enhancements. Many seek minor modifications to administrative requirements under various financing programs, especially TIFIA. At least one application (for initial projects under the Oregon Innovative Partnerships Program) acknowledges the importance to freight movements of certain roads proposed for improvement, but it does not mention potential air quality benefits. Like most other Federal financing tools, SEP-15 appears to be focused primarily on large capacity enhancement projects. Since capacity enhancements for freight may have ancillary air quality benefits, planners should stress such benefits in SEP-15 applications and note any extra benefits that would accrue if the project were accelerated.

4.4 EPA National Clean Diesel Campaign

In addition to the traditional (U.S. DOT) funding and financing programs, the U.S. EPA also provides Federal funding for freight air quality as part of its National Clean Diesel Campaign (NCDC). The programs administered by the EPA under the NCDC are described in this section and summarized in Table 4.3. Funding levels are described in terms of the program's normal annual appropriation (in this case, for FY 2009) and in terms of one-time funding from the American Recovery and Reinvestment Act of 2009 (ARRA). ARRA was a $787 billion fiscal stimulus package designed as a response to the economic crisis. Among other things, the bill provided grant funding for programs and projects that have environmental benefits, including those that relate to freight emissions.

---

[85] 23 U.S.C. 502.

Table 4.3 EPA Funding Programs

Funding Program	Eligibility	Funding Level (FY 2009)	Freight Air Quality Application	Project Size
National Clean Diesel Funding Assistance Program	Provides grant money to reduce diesel emissions through technology applications, alternative fuels, and equipment replacement. Technologies must be certified by EPA and CARB.	EPA: $32 million ARRA: $156 million	• Add-on emissions control devices • Idle reduction devices • Alternative fuels • Shore-power systems • Engine repowers/upgrades • Equipment replacement	Small. Total FY 2009 funding is $32 million, but American Recovery and Reinvestment Act provided substantial one-time resources.
Clean Diesel Emerging Technologies Program	Provides grant funds for the implementation of emerging technologies (not yet certified by CARB or EPA) to reduce diesel emissions.	EPA: $4 million ARRA: $20 million	• Certain emissions upgrades to previously manufactured engines • List includes selective catalytic reduction systems, diesel oxidation catalysts, and emissions upgrade kits	Small. Total FY 2009 funding is $4 million, but American Recovery and Reinvestment Act provided substantial one-time resources.
SmartWay Clean Diesel Finance Program	Provides nonprofits with grant money to establish low-cost financing programs for buyers of eligible diesel vehicles, equipment, and emissions control retrofit devices	EPA: $6 million ARRA: $30 million	• Exhaust controls • Engine upgrades • Aerodynamic improvements • Low-rolling resistance tires • Purchase of new equipment meeting stricter emissions requirements • Idle reduction devices	Small. Total FY 2009 funding is $6 million, but American Recovery and Reinvestment Act provided substantial one-time resources.
State Clean Diesel Grant Program	Formula to states for the implementation of clean diesel grant and loan programs.	EPA: $18 million ARRA: $88 million	• EPA- or CARB-certified engine retrofits • EPA-certified idle reduction technology • Technologies from EPA's Emerging Technologies List • Incremental costs of early replacement/repower with certified engine configurations	Small. Total FY 2009 funding is $18 million, but American Recovery and Reinvestment Act provided substantial one-time resources.

The NCDC was established to promote diesel emission reduction strategies. There are two main components of the program:

- **A regulatory program** that includes new fuel and emissions standards designed to reduce emissions from new (and in some cases remanufactured) diesel engines; and

- **The Diesel Emissions Reduction Act (DERA),** which is a set of innovative partnership strategies that seeks to encourage government, nonprofit, and industry stakeholders to adopt policies that reduce their fleet's emissions. In FY 2009, total DERA funding was $60 million. However, the ARRA provided the DERA program with an additional $300 million in one-time funding on top of the existing program's annual appropriations.

There are four distinct programs within the DERA. Each is described below.

- **National Clean Diesel Funding Assistance Program.** This program provides grant funding to reduce emissions from existing diesel engines through a variety of strategies, such as: add-on emission control technologies; idle reduction devices; alternative fuels; engine repowers; engine upgrades; and/or vehicle or equipment replacement. Funds also may be used to create innovative finance programs to fund diesel emissions reduction projects. Only technologies that have been verified and certified by the EPA and California Air Resources Board (CARB) may be implemented with these funds.

- **Clean Diesel Emerging Technologies Program.** This program is similar to the National Clean Diesel Funding Assistance Program, except that it is limited to the implementation of "emerging technologies," defined as devices or systems that reduce emissions from diesel engine powered vehicles or equipment that has not been certified or verified by EPA or the CARB, but for which an approvable application and test plan have been submitted for verification. A list of such technologies is available at: www.epa.gov/cleandiesel/prgemerglist.htm.

- **SmartWay Clean Diesel Finance Program.** SmartWay is a partnership with the freight sector that seeks to encourage private sector adoption of various emissions reduction strategies. The program is predicated on the notion that the improvements will pay for themselves through fuel savings and/or productivity enhancements. Through nonprofit partners, SmartWay offers low-cost financing to truckers and fleet owners interested in installing emission reduction technologies on their trucks. Eligible applications include exhaust controls, engine upgrades, idle reduction devices, aerodynamic improvements, and vehicle or equipment replacement, among other things. Upgrades must be EPA and CARB certified.

- **State Clean Diesel Grant Program.** Through this program, EPA allocated formula funding to states to establish grant and loan programs for clean diesel projects. Participating states receive two-thirds of program funds as base funding; those that match the entire base funding amount are awarded extra funds equal to one-half of their base funding. The funds may be used to implement EPA- or CARB-certified engine retrofits, EPA-verified idle reduction technologies, technologies from EPA's Emerging Technologies List, and the incremental costs associated with early engine replacement/repowering with certified engine configurations.

DERA grant and loan funding opportunities are available either at the national level or through a network of regional diesel collaboratives, which partner with the EPA to implement the various emissions reduction strategies in their areas. The regional collaboratives are as follows:

- Northeast Diesel Collaborative (Maine, Vermont, New Hampshire, Massachusetts, Connecticut, Rhode Island, New York, New Jersey, Puerto Rico, Virgin Islands);

- West Coast Collaborative (Washington, Oregon, Idaho, California, Nevada, Arizona, Hawaii, Alaska);

- Mid-Atlantic Diesel Collaborative (Pennsylvania, West Virginia, Virginia, District of Columbia);

- Midwest Clean Diesel Initiative (Minnesota, Michigan, Wisconsin, Illinois, Indiana, Ohio);

- Southeast Diesel Collaborative (Kentucky, Tennessee, North Carolina, South Carolina, Mississippi, Alabama, Georgia, Florida);

- Blue Skyways Collaborative (New Mexico, Texas, Louisiana, Arkansas, Oklahoma, Kansas, Nebraska, Iowa, Minnesota); and

- Rocky Mountain Diesel Collaborative (Colorado, Utah, Wyoming, Montana, South Dakota, North Dakota).

Many regional diesel collaboratives also have established their own funding and financing initiatives. Practitioners should, therefore, check with the regional collaborative for their state to find out about potential funding opportunities. This is especially important when considering projects for Recovery Act funding, as application windows for those funds already may have closed. More information on Clean Diesel Campaign funds can be found on the EPA website, at http://epa.gov/otaq/diesel/grantfund.htm.

4.5   STATE, LOCAL, AND NONPROFIT PROGRAMS

Some states, localities, and regions have their own diesel emissions reduction programs. Like their national counterparts, these programs employ a variety of strategies to encourage the adoption of clean diesel technologies. There also are some nonprofit organizations and public-private partnerships that work to reduce diesel emissions through advocacy and outreach as well as offering low-cost financing for retrofits. Several of these programs are described in this section. While this does not represent a comprehensive list of all such programs across the country, it can serve as a reference of the types of programs that may be available for state or local project sponsors. Planners should check with their state environmental agencies and advocacy groups to get information on programs for their state.

4.5.1   STATE DIESEL EMISSION PROGRAMS

Examples of state diesel emissions reduction grant programs in California, Oregon, and Texas are described below.

Carl Moyer Program (California). The Carl Moyer Memorial Air Quality Standards Attainment Program is a state and local partnership administered by the California Air Resources Board (CARB), itself a division of the California EPA. The program provides incentive grants to encourage the adoption of cleaner-than-required engines and equipment. The grants can be used for on road, off-road, stationary, marine, and locomotive engines, as well as fleet modernization efforts and idle reduction technologies. Since its inception in 1998, the program has provided over $154 million in grant awards to California-based private companies.

Oregon Clean Diesel Initiative. The Oregon Department of Environmental Quality offers grants and tax credits to qualifying businesses that retrofit their diesel engines with emissions reduction equipment that reduces diesel particulate matter by at least 25 percent. To qualify for incentives, the retrofit technology must be verified by the U.S. EPA and/or the California

Air Resources Board (CARB). Alternatively, the Oregon DEQ can determine that the device has been through comparable testing. The device also must be used in Oregon for at least three years following the retrofit, for at least one-half of the total miles driven or hours operated. Grants are awarded to entities based on several preferences, including:

- Proportion of miles driven or hours operated in Oregon;

- Benefit to disadvantaged populations or areas that already have high concentrations of particulate matter;

- High cost-effectiveness;

- Commit funding, materials, or expertise from third parties;

- Reduce more emissions in Oregon;

- Applicants demonstrate a commitment to making additional air quality improvements; and

- Applicants have the capacity to complete the project.

Any person may apply for a tax credit in Oregon after completing a qualifying diesel retrofit. The credit can reduce the applicant's Oregon tax liability by up to one-half the cost of retrofitting the diesel engine.

Texas Emissions Reduction Plan (TERP). TERP is a comprehensive set of incentive programs aimed at improving air quality in Texas. Through the Texas Commission on Environmental Quality (TCEQ), TERP provides grant money to eligible projects to reduce diesel emissions (primarily NO_x) from high-emitting mobile and stationary sources, including trucks and locomotives. Only projects located in nonattainment areas and the counties adjacent to them are eligible. This includes the Houston-Galveston-Brazoria, Dallas-Fort Worth, Beaumont-Port Arthur, Tyler-Longview, Austin, and San Antonio areas. There are two programs housed within the TERP:

- **Emissions Reduction Incentive Grants** help offset the costs of reducing emissions from high-emitting diesel engines. Eligible projects include purchase or lease, repowering, replacement, or retrofits of diesel locomotives, trucks, stationary equipment, and marine vessels. Also eligible are refueling structures (for qualifying fuel), rail relocation and improvement, and electrification/idle reduction efforts (either on-site or on-vehicle).

- **Rebate Grants** are only issued for repowering or replacement of certain non-road and on-road vehicles and equipment. The equipment must be operated 75 percent of the time in the eligible counties and must have a service life of five or seven years. The application process is designed to be shorter and less burdensome, which makes it more attractive for smaller businesses that might otherwise be put off by a complicated application process. In fact, part of the program funding is specifically set aside for qualifying small businesses.

4.5.2 LOCAL AND REGIONAL PROGRAMS

There also are a number of local and regional programs around the country. These tend to be located in areas with significant air quality problems. Two examples are described below for the Los Angeles region and the Houston-Galveston area. Both programs include elements that respond specifically to freight-generated emissions.

Gateway Cities Clean Air Program. The Gateway Cities Clean Air Program was created to provide a financial incentive to help reduce air pollution in Southern California. It was a six-year pilot program that began in 2002 and ended in 2008. The program was managed by the Gateway Cities Council of Governments (an intergovernmental organization for 27 cities in the Los Angeles/Long Beach area, with a combined population of two million people) with funding from the Port of Long Beach, Port of Los Angeles, California Air Resources Board, South Coast Air Quality Management District, and United States EPA. Through the Gateway Cities Clean Air Program, participating truck owners received $24.5 million in grants to replace 643 highly polluting older model heavy-duty diesel trucks with newer and cleaner, lower-emitting trucks.

Houston-Galveston Clean Vehicles Program. The Houston-Galveston Area Council (H-GAC) is the designated MPO for the Houston-Galveston region. Like Southern California, the Houston area faces significant air quality problems, stemming in part from growing freight activity at the Port of Houston and around the region. To help combat the problem, H-GAC has implemented a Clean Vehicles Program, which provides grants to fund projects that improve air quality. Engine retrofits, repowering/replacement, alternative fuels conversion, and the establishment of publicly accessible alternative fuels infrastructure are all eligible. Public or private fleets operating primarily within the eight-county ozone nonattainment area can apply for funding. The vehicles must spend at least 75 percent of their operating hours in the eight-county region and must travel more than 12,000 miles per year.

For private fleets, H-GAC will reimburse project sponsors at a rate of $70,000 per ton of NO_x reduced per year, up to 75 percent of total costs; for public fleets, the rate is $150,000 per ton of NO_x reduced per year, also up to 75 percent of project costs. H-GAC staff evaluate applications based on the expected annual emissions reduction and tons per year reduced; and capital cost-effectiveness, which is the cost per ton of the emissions reduction.

Once a project is awarded grant funding, the sponsor implements it and invoices H-GAC for reimbursement of project costs. Participants are required to submit quarterly monitoring reports for a period of five years, and H-GAC staff are authorized to audit and/or visit the project to ensure compliance.

4.5.3 Cascade Sierra Solutions

Cascade Sierra Solutions (CSS) is a multistate, corridor-level nonprofit initiative that seeks to reduce freight emissions by helping truck owner-operators and fleets make technological improvements that will save fuel and reduce diesel emissions. The program covers Washington, Oregon, and California. It provides a variety of services, including:

- **Regulatory advice,** such as information about idling/emission rules and air quality goals for states and local governments;

- **Equipment selection,** providing information about different brands and models of emission reduction equipment;

- **Financing,** or matching customers with financing options allowing them to purchase and install the devices;

- **Installation contracting,** which involves coordinating the installation of fuel-saving technology by qualified contractors; and

- **Monitoring, testing, certification, and reporting,** whereby CSS monitors the use and operation of the devices to ensure that program objectives are being met.

In 2008, CSS received a $1.13 million grant through the EPA SmartWay partnership to implement a lease program with the goal of installing emission and idle reduction technology on 1,700 trucks nationwide. The CSS program (called Everybody Wins USA) offers truck owners interest rates of eight to 11 percent, a three-year repayment period, and the ability to purchase the equipment for $10 at the end of the term.

5.0   Case Studies

This section presents a series of case studies of freight projects and programs that seek to improve air quality and reduce freight-related emissions. These case studies provide real-world examples of the operational, infrastructure, and technology solutions being used to solve freight air quality problems. Each case study identifies "key themes" for freight and air quality practice, allowing practitioners to quickly identify case studies that may be most relevant to their interests. The case studies are organized according as to whether the emission reduction strategies employed are technological in nature (e.g., diesel engine retrofits) or operational (e.g., congestion management).

5.1   TECHNOLOGICAL FOCUS

5.1.1   NORTH CENTRAL TEXAS COUNCIL OF GOVERNMENTS DIESEL FREIGHT VEHICLE IDLE REDUCTION PROGRAM

The North Central Texas Council of Governments (NCTCOG) is an association of local governments in the Dallas-Fort Worth (DFW) region. NCTCOG serves 16 counties centered around Dallas and Fort Worth, and is comprised of over 230 member governments, including counties, cities, school districts, and special districts. NCTCOG also is the Federally designated MPO for the Dallas-Fort Worth Metroplex. The MPO planning area includes all of Dallas, Collin, Denton, Tarrant, and Rockwall counties and parts of Ellis, Johnson, Kaufman, and Parker counties, for a total area of over 5,000 square miles. Of the 6.6 million people that live in the 16-county area, 92 percent (6.1 million) reside in the MPO planning area.

KEY THEMES
• Truck stop electrification
• APUs
• Hybrid trucks
• Multijurisdictional coordination (MPO/local)
• Innovative use of CMAQ funds on private facilities

All nine of the MPO counties were designated as ozone nonattainment areas by the EPA in 2004. Recognizing the impact that truck and rail freight operations have on regional air quality, the NCTCOG created the Diesel Freight Vehicle Idle Reduction Program as a way to link city and regional policies and investments related to idle reduction programs. The program seeks the use of Federal funds in order to improve the area's goods movement infrastructure and generate air quality benefits. Through the Regional Transportation Council (the MPO's policy-making body), NCTCOG offers grant funding for projects that reduce unnecessary truck idling, thereby also reducing NO_x emissions. The MPO has partnered with several stakeholders, including TxDOT, local governments, and private sector businesses to identify and fund specific freight system improvements or technological upgrades to mitigate freight-related vehicle emissions. The primary focus of the program is on truck stop electrification and anti-idling efforts.

Each year, NCTCOG issues a call for projects outlining general program guidelines, eligibility, and application procedures. Primary consideration is given to projects that achieve emissions reductions mostly within the nonattainment counties, but consideration is given to projects that reduce emissions in the EPA Blue Skyways Collaborative states, focusing on the I-35 corridor. Federal Congestion Mitigation and Air Quality Improvement Program (CMAQ) funds have been used under this program to make improvements to private freight facilities under a section of the program that allows public money to be spent on private freight facilities if a public benefit can be demonstrated. Grant funding also comes from EPA money. NCTCOG also makes use of EPA programs, including the National Clean Diesel Funding Assistance Program, the National Clean Diesel Emerging Technology Program, and the SmartWay Clean Diesel Finance Program, all of which are under DERA.

During the 2008 round, four projects were awarded funding totaling nearly $746,000 to fund on-board idle reduction and truck stop electrification projects, three of which were sponsored by private sector entities:

- **Craufurd Manufacturing** was awarded nearly $612,000 for a truck stop electrification project. The company makes AireDock technology, which is recognized by the EPA as an effective idle-reduction application. The project involves outfitting 80 truck parking spaces at the Star Travel Plaza on I-35 in Denton with AireDock systems. MPO staff estimate that the project will reduce daily NO_x emissions by 0.0898 tons, or 328 tons over the life of the project. Funds for the project came from the CMAQ program.

- **Summit Transportation** received $64,000 of EPA funds to install APU on its fleet of 16 trucks. The company is based in the DFW region, and 90 percent of its cargo is loaded and unloaded in the Metroplex. According to the company, the average fleet idle time is 38 percent. The APU will help to reduce time spent unnecessarily idling. The company also participates in the EPA SmartWay Transport Partnership. It was estimated that the APU would reduce NO_x emissions by 7.41 tons over the life of the project.

- **The City of Fort Worth** received a $50,000 grant to purchase two hybrid trucks to be used for traffic light maintenance and excavation. The new trucks will replace two conventional trucks, both of which must idle when used at a work site for power take-off (PTO) applications. The project is being funded through CMAQ; staff estimate that the new trucks will reduce NO_x emissions by a total of about two tons over the project life.

- **Southeastern Freight Lines,** a trucking firm based in South Carolina, maintains six terminals along the I-35 corridor, including a Dallas location with a fleet of 150 trucks. The company received $20,000 to install APU on four of its Dallas-based sleeper cab trucks, which will reduce NO_x emissions by about 2.5 tons over the life of the project. Southeastern Freight Lines also participates in SmartWay, as well as the Southeast Diesel Collaborative.

NCTCOG offers application assistance to prospective applicants, including a spreadsheet-based idle reduction calculator that estimates NO_x reduction and cost-effectiveness in both the DFW nonattainment area and the Blue Skyways Collaborative region.

5.1.2 CASCADE SIERRA SOLUTIONS

Cascade Sierra Solutions (CSS) is similar to the NCTCOG Diesel Freight Vehicle Idle Reduction Program, but it has been implemented at a multi-state corridor level, in response to the multijurisdictional nature of freight movements. The program focuses on the states of Washington, Oregon, and California but has provided advice and financing for truck owners nationwide. CSS seeks to reduce freight emissions by helping truck owner-operators and fleets make technological improvements that will save fuel and reduce diesel emissions. CSS helps truck owners find financing for fuel-saving technology applications like APU as well as new trucks; it also helps them process tax credit applications with various other government entities that offer incentives for emissions reduction upgrades. CSS obtains operational funding from a combination of donations, grants from private foundations, limited client service fees, and various government grant programs. CMAQ and DERA are two key sources of Federal funding for CSS. The organization also receives grant money from various state and local environmental agencies.

> **KEY THEMES**
> - Truck replacement
> - Diesel particulate filters (DPF)
> - APUs
> - Multijurisdictional coordination (state to state)
> - Effective public/private coordination

A few recent CSS projects are outlined below.

- **Mesilla Valley Transportation** is a Las Cruces, New Mexico-based trucking firm that specializes in transporting time-sensitive goods between manufacturing centers in the United States and along the Canadian and Mexican borders. CSS helped the company obtain energy tax credits that enabled them to equip 300 truck tractors with APU. The company also has been actively greening its truck fleet for some time by upgrading to new fuel-efficient equipment and installing various fuel-saving aerodynamic equipment such as low-rolling resistance tires, trailer side skirts, and aluminum wheels. All told, these improvements have improved fuel economy by more than 30 percent. Each year, the company saves over 5,000 gallons of fuel per truck while eliminating 60 tons of greenhouse gas emissions.

- **Cross Creek Trucking** is an Oregon-based carrier that specializes in long-haul less-than-truckload (LTL) service, primarily for agricultural goods but also for other types of freight. Long-haul trucks spend a large amount of time at idle, either parked en route to a destination or staged for pickup or delivery. In fact, the company determined that each of the 115 trucks in its fleet was idling for 10 to 13 hours per day. Burning one gallon of diesel fuel per hour, Cross Creek was paying $650,000 per year just to idle its fleet. CSS helped the company to secure tax credits from the Oregon Department of Environmental Quality along with low-interest financing to equip its fleet with APU. This reduced Cross Creek's annual idling cost to $110,000. Fleet fuel efficiency has increased by more than 18 percent. Cross Creek currently is working on other fleet improvements that will save fuel, reduce emissions, and keep the company in compliance with new emissions laws.

- **Devine Intermodal** offers a variety of intermodal transportation services in Northern California and Nevada. Devine employs both company drivers and independent owner-operators. Contracted drivers often have a difficult time upgrading their equipment because they typically do not have the financial resources to do so. To help overcome this obstacle, CSS worked with Devine's management to apply for and receive truck replacement grants from the State of California and develop a low-cost financing program for independent drivers. The company offered loan guarantees to prospective truck buyers to improve their credit scores and give them access to below-market financing. At a meeting hosted by CSS at its Sacramento Outreach Center, company officials presented the plan to owner-operators and CSS staff processed loan prequalifications on the spot. In all, 64 independent drivers obtained truck replacement grants and financing.

- **Bettendorf Trucking and Joe Costa Trucking,** two firms that are jointly owned and serve the forest products industry in Oregon and California, have historically focused on staying profitable by maximizing operating efficiency. Part of this strategy has been to minimize fleet costs by keeping the current fleet running as long as possible, rather than purchasing new trucks. Although this makes good business sense, these trucks are older and emit more than comparable newer models. These older trucks also are one of the main targets of California's strict new emissions rules. Working with CSS, the companies received grants through California's Carl Moyer Memorial program to install diesel particulate filters on 49 trucks, reducing exhaust emissions by 85 percent. The companies also have received 21 truck replacement grants worth $50,000 each through the California Proposition 1B program.[86]

---

[86] Proposition 1B was a transportation bond package passed by California voters in 2006. It gave the State the authority to sell about $20 billion in bonds for transportation projects, $3.2 billion of which was devoted to goods movement and air quality.

5.1.3 SAN PEDRO BAY PORTS CLEAN AIR ACTION PLAN

The Ports of Los Angeles and Long Beach lie adjacent to each other about 30 miles south of downtown Los Angeles. They are the two busiest container ports in the United States and, taken together, the fifth busiest in the world. More than $260 billion worth of goods move through the ports each year. Southern California has well-documented air quality issues, with several counties or portions of counties in the region in nonattainment status for ozone, PM_{10}, and $PM_{2.5}$. Although the region's air quality problems stem from a variety of sources, the ports are a significant source of diesel emissions in the area, particularly given the explosive growth in international trade that has occurred over the last few decades.

The San Pedro Bay Ports Clean Air Action Plan (CAAP) is an emissions reduction plan adopted by the ports to improve air quality in the Los Angeles basin by implementing strategies to reduce port-related emissions from ships, trains, trucks, terminal equipment, and harbor craft. It represents the culmination of a series of air quality initiatives that began with a promise from the City of Los Angeles to have no new emissions. Although the San Pedro Bay ports are a significant source of diesel emissions in the region, the Federal and state governments have no authority over many port emissions sources (such as foreign-flagged vessels), so the ports can employ emissions strategies that other entities cannot.

The CAAP is a collaborative program that has been endorsed and adopted by both ports. Getting both ports on-board was critical because port tenants need assurance that they would not be subject to different requirements at each port. The plan also has the support of the South Coast Air Quality Management District, EPA, and California Air Resources Board. The CAAP is a five-year plan, but it also has a long-term component that describes how five-year emissions reduction actions would be integrated into port operations over the long term, and their expected impact on emissions. The plan targets PM emissions, but SO_x and NO_x reductions are secondary goals.

The CAAP focuses on three implementation strategies:

- **Tenant Leases.** Whenever new development occurs on port property, the port works with the tenant to put mitigation measures into the lease. When port tenants amend or renew their leases, the port must comply with the CEQA and the NEPA (if applicable). The CAAP serves as the guiding document for developing mitigation strategies during the Environmental Impact Statement (EIS) phase of the project. The port then negotiates with the tenant to incorporate feasible mitigation measures into the lease. Measures that are not feasible for the

> **KEY THEMES**
> - Truck replacement
> - Retrofits
> - Alternative fuel
> - Use of incentives to encourage private sector implementation of air quality improvement measures
> - Incorporation of mitigation measures into new port development

tenant to undertake become the responsibility of the port. Examples of mitigation measures placed in a lease include requiring the use of shore power by ships berthed at the terminal and tenant adoption of clean yard equipment.

- **Incentives.** The ports provide monetary incentives to retrofit older trucks accessing the terminals with emissions control devices, or replace them with new, cleaner models. This approach has been effective since trucks often access multiple terminals and are outside of the control of any one tenant. There also are incentives to use ULSD and to reduce vessel speeds when approaching the ports.

- **Tariffs.** The CAAP calls for tariff changes to encourage the adoption of emissions reduction strategies by vessels calling on the port, but to date these have not been implemented largely because of the economic downturn.

Although the plan considers five emissions source categories (vessels, harbor craft, cargo handling equipment, rail, and truck), the ports have determined that focusing on trucks offers the best "bang for the buck"; accordingly, they have concentrated significant resources on the Clean Trucks Program. This program seeks to replace or retrofit 16,000 harbor trucks in five years by:

- Barring older, more polluting trucks from entering the port terminals;

- Issuing grants to replace or retrofit trucks to reduce emissions; and

- Collecting "Truck Impact Fees" from noncompliant trucks to support the Clean Truck Program.

As of October 2008, pre-1989 trucks were banned from the terminals. Model year 1989 and later trucks will be progressively banned until only 2007 and newer models (which employ the cleanest diesel technology) are permitted access to the terminals without paying the impact fee. Collection of the truck impact fee began in February 2009. Beneficial cargo owners are now charged $35 per TEU to access the terminals.[87] This fee is collected by PortCheck, a nonprofit sister company to PierPass (described below). The ports are using the proceeds to subsidize new truck purchases. In addition, about 50 grants have been awarded to retrofit existing trucks, replace them with newer models, or adopt alternative fuels such as LNG.

---

[87] 'Beneficial cargo owner' means the importer of record who takes possession of a shipment at the final destination (i.e., not a third-party freight carrier).

The ports provide most of the program funding, but they are getting some state support for the Clean Trucks Program. Overall, the ports have proposed to allocate $200 million to the Clean Trucks Program alone.

5.1.4 OREGON DEPARTMENT OF ENVIRONMENTAL QUALITY COLUMBIA RIVER BARGE TOW REPOWER

The Columbia River cuts through the Cascade Mountains, virtually at sea level, making it an important river connection between the American Midwest and the West Coast. In fact, the Columbia and Snake River system is the largest inland waterway west of the Mississippi River. Large volumes of agricultural products traverse the Columbia via barge en route to export markets in Asia. These rivers are the largest wheat export system in the United States with exports on the lower Columbia River projected to double by 2025. Geographic barriers and climatic characteristics make the Columbia River Gorge particularly susceptible to air pollution problems stemming from several sources, including highway, rail, and barge transportation; industrial activities; and major population centers.

> **KEY THEMES**
> - Diesel retrofit and repower
> - Use of monetary/tax incentives to encourage private sector air quality improvement measures
> - Use of multiple Federal and state funding sources to accelerate a freight air quality project

Shaver Transportation Company began providing marine freight transport services on the Willamette River in the 1880s with steam powered sternwheelers. With the introduction of diesel engines and improved navigation, the company extended its service area to the Columbia and Snake rivers. The company currently provides ship assist services in the Portland/Vancouver harbor and upriver barge service as far as Lewiston, Idaho.

Like locomotives, barge tow engines have exceptionally long service lives. Normal business practice in the barge tow industry is to overhaul engines periodically, which is much cheaper than purchasing new engines, despite the expected fuel savings. In this case, to achieve greater air quality benefits Shaver instead proposed to replace the two 33-year old diesel engines on its tug CASCADES with EPA Tier 2-compliant MTU 12V4000 M60 engines, which emit no more than the certified limits for NO_x and PM.

These new engines have the same power output as the old ones but are much more fuel efficient, saving the company about 158,000 gallons of diesel fuel per year while eliminating 1,600 metric tons per year of carbon dioxide. The new engines reduced the barge's NO_x emissions by 69 percent, or about 209 tons annually. Although data for PM emissions from the old engines are unavailable since they predate any emissions regulations, other reports indicate that PM emission reductions somewhere between 60 and 85 percent can be expected from the repowered engines. To achieve equivalent results (in both NO_x and PM) from heavy-duty trucks would require retrofitting about 250 trucks at a cost of over $3.7 million.

The overall project cost for the engine repower was $1.9 million with 35 percent of that coming from Federal grant funds and state tax credits. The total cost included not only the engines themselves but other integrally related costs like gear box reductions, keel engine coolers, shipyard costs, and engineering. The Oregon Department of Environmental Quality secured a $100,000 DERA grant for the project. The project also received two tax credits offered by the State of Oregon – one for $220,000 through the Business Energy Tax Credit program administered by the Oregon Department of Energy, and an ODEQ Diesel Repower Tax Credit for $350,215. The Business Energy and Clean Diesel Tax Credits reduce Oregon taxpayer liability for cleaner and/or more efficient diesel engine repowers by covering a portion of the incremental costs of qualifying projects. The remaining funding (about $1.2 million) was provided by Shaver Transportation.

5.1.5 Chicago CMAQ Locomotive Purchase

> **Key Themes**
>
> - Locomotive repowering
> - Innovative local matching arrangement
> - Use of CMAQ funds on private rail equipment

The Chicago region experiences air quality issues associated with a large (and growing) population, increasing vehicle-miles traveled (VMT), and a concentration of industrial activities. All or part of eight counties in Northeastern Illinois are in a designated nonattainment area for ozone and PM2.5. Chicago's status as the nation's premier rail freight hub (all seven Class I railroads connect there) means diesel emissions from rail locomotives are a pressing concern. This is especially true near rail yards, since many of the switcher locomotives used to reposition rail cars and assemble trains are older and lack sophisticated emissions control technology.

The Chicago Metropolitan Agency for Planning (CMAP, the MPO for the region) has a CMAQ project selection committee which evaluates projects applying for funding through the area's CMAQ apportionments. Each year, there is an open call for projects. In 2007, the City of Riverdale (located about 20 miles south of downtown Chicago) approached the committee with a project concept focused on replacing switcher locomotives at a large rail yard owned by CSX. After consultations between officials from the MPO and the City, as well as CSX, an application for funding was submitted. The project then was selected for a CMAQ grant.[88] The funds were used to replace five switchers in the CSX Riverdale yard with cleaner Genset locomotives.

CMAQ normally requires a 20 percent local match.[89] Historically, local agencies acting as sponsors of air quality projects have provided the

---

[88] The MPO evaluates all funding applications and performs emissions modeling to estimate air quality impacts. Projects are rated based on the cost per kilogram of pollutant(s) reduced.

[89] However, the Energy Independence and Security Act of 2007 provided for up to a 100 percent Federal share for CMAQ projects in FY 2008 and 2009.

match. In this case, however, the CSX railroad provided the match, by paying for the fifth locomotive. The CMAQ grant was used to purchase the other four gensets. Under the terms of the funding agreement, CSX is required to keep the new locomotives within the nonattainment area for at least 10 years. This does not present a problem for switch locomotives since they operate only at the switchyard.

The successful implementation of this project garnered the attention of other railroads in the Chicago region, and in subsequent years the MPO has received several applications for similar projects. Demand has been so great, in fact, that the match required of the railroads was recently raised to 35 percent. It was determined that the fuel savings the railroads realize through adopting cleaner locomotives more than offsets the higher match requirement. In FY 2009, five locomotive retrofit projects were awarded funding totaling nearly $11 million.[90]

5.2   OPERATIONAL FOCUS

5.2.1   PORTS OF LOS ANGELIS/LONG BEACH PIERPASS/OFFPEAK PROGRAM

PierPass is a nonprofit entity created by marine terminal operators at the San Pedro Bay ports in Southern California. It was developed in 2004 in response to a bill introduced in the California legislature that would have imposed a "peak-hour surcharge" on all containers entering or exiting the Ports of Los Angeles and Long Beach between the hours of 8:00 a.m. and 5:00 p.m. Faced with the prospect of having a tax imposed on daytime container movements, the terminal operators banded together and proposed a private sector solution. Although congestion relief is the primary focus of the program, it also has important air quality benefits.

The program (known as OffPeak) provides incentives for shippers to move cargo at night and on weekends, rather than during congested daytime hours. The incentive is in the form of a "traffic mitigation fee" imposed on all cargo imported and exported through the ports. The fee currently is set at $50 per 20-foot equivalent unit (TEU),[91] but can vary according to business conditions and the actual costs of running the OffPeak program. Cargo owners then receive a refund for all loads that are handled

[90] Chicago Metropolitan Agency for Planning, 'CMAQ Multi-Year Program for Northeastern Illinois – FY 2009', retrieved from http://www.cmap.illinois.gov/ uploadedFiles/committees/cmaq/documents/fy09/Approved%20FY%20 2009%20CMAQ%20Program.pdf.

[91] One TEU corresponds to a standard 20-foot shipping container. The most common containers are 40 feet long, or two TEUs.

during off-peak hours, which are defined as 6:00 p.m. to 3:00 a.m. on Monday through Thursday, and 8:00 a.m. to 6:00 p.m. on Saturday. This reduces congestion at the ports' gates and takes trucks off the road during the busiest hours of the day, both of which reduce emissions from idling traffic. Intermodal cargo that already is being charged a fee by the Alameda Corridor Transit Authority (ACTA) is exempted from the charge. Revenues from the traffic mitigation fee are used to support the operating costs associated with maintaining extended port gate hours.

OffPeak usage of the marine terminals has grown fairly steadily since the program was implemented in late 2005, and was approaching 40 percent by the end of 2008.[92] It is estimated that about 68,000 truck trips each week occur during OffPeak shifts, and in three years of operation, OffPeak removed more than nine million truck trips from local freeways during peak commuting hours.[93] Since trucks represent a large share of the traffic mix on area freeways (especially near the ports), this shift to off-peak trip making leads directly to fewer trucks idling in traffic during the day, wasting fuel and increasing emissions. It also means shorter queues at the port gates, reducing the amount of time trucks must idle while waiting to access the terminals.

One key strength of OffPeak is its responsiveness to industry needs. The program was designed to be flexible so that fees and OffPeak terminal operations could be adjusted according to business trends and cost factors. For instance, the marine terminal operators recently announced that one OffPeak shift would be eliminated because of the large drop in cargo volumes associated with the current economic downturn. When volumes begin to grow again, shifts can be added to cope with the increased demand. The program also has benefited truckers who access the ports, two-thirds of whom have a positive opinion of OffPeak. Truckers report increased income (from being able to make more trips during a work shift), reduced congestion, and lifestyle benefits associated with more flexible work schedules.[94]

---

[92] http://www.pierpass.org/.

[93] PierPass, "PierPass OffPeak program diverts more than nine million truck trips from daytime traffic over first three years of operation," press release dated July 23, 2008.

[94] Fairbank, Maslin, Maullin and Associates, *PierPass Los Angeles/Long Beach Harbor Truckers Survey,* 2006.

5.2.2 PORT OF SEATTLE SR 519 INTERMODAL ACCESS PROJECT

The Port of Seattle is a major marine freight gateway for the Pacific Northwest, particularly for Asian trade. In 2007, the port was ranked as the nation's seventh busiest.[95] SR 519 is a key truck access route to the port. In recent years, railroad grade crossings and increasing freight and passenger traffic at the Port of Seattle have led to congestion around the port complex. The surrounding SoDo district also experiences heavy pedestrian traffic from the many entertainment and tourist attractions in the area, leading to concerns about safety and conflicts between pedestrians and motorized vehicles.

In response, the Washington State DOT (WSDOT) has been implementing a series of improvements aimed at streamlining traffic in the area. Phase 1 of the project consisted of the construction of a new overpass between Occidental Avenue South and I-90, separating truck, automobile, and pedestrian traffic from the railroad tracks. In Phase 2, WSDOT will connect a westbound off-ramp from I-5 and I-90 to the existing South Atlantic Street overpass, make key intersection improvements, and build a bridge on South Brougham Way over the railroad tracks. Construction of Phase 2 began in October 2008.

The proposed improvements to SR 519 will separate car, freight, pedestrian, and rail traffic around the port, thereby improving passenger, freight, and pedestrian mobility around the port complex. The Environmental Assessment for Phase 2 found that implementation of the project would improve air quality by reducing congestion and idling times for freight trucks and trains as well as passenger traffic. The project will comply with NAAQS, the Washington State Implementation Plan (SIP) for carbon monoxide, and all requirements of the Washington Clean Air Act and the Federal Clean Air Act.

Like many recent freight projects, the SR 519 improvements are being financed through a public/private partnership involving Federal, state, local, and private funds. Phase 1 of this project cost $109.3 million. This money came from a variety of sources, as outlined below.

- **State preexisting funds and Freight Mobility Strategic Investment Board (FMSIB)** funds provided $30.9 million. The FMSIB is a state agency created 10 years ago and charged with creating a comprehensive, coordinated state program to enhance freight mobility between and among local, national, and international markets, thereby improving trade opportunities. A corollary mission is to lessen the impact of freight movements on local communities. The Board accomplishes

KEY THEMES

- Rail/highway grade separation
- Port access improvements
- Use of air quality benefits to help move a highway port access project forward
- Effective use of public/ private partnership with multiple funding sources

---

[95] Port of Seattle.

these goals by advocating for freight in project planning and leveraging funding from various public and private sources to move freight projects forward.

- **The Federal Highway Administration** provided $54.6 million through the Surface Transportation Program, National Highway System program, and some demonstration funding.

- **The remaining $23.8 million came from various other sources,** including the Port of Seattle, the BNSF Railroad, the City of Seattle, King County, the Federal Transit Authority, and a local Public Facilities District.

Phase 2 is expected to cost $84.4 million, which also was assembled from several sources:

- **The State Transportation 2003 Account** has committed $72.9 million from the "Nickel Funding" package. This package was adopted by the State Legislature in 2003 to finance 158 transportation projects over a 10-year period, and is funded through a five cent per gallon gas tax increase, a 15 percent increase on gross weight fees for heavy trucks, and a 0.3 percent increase on the sales tax for motor vehicles;

- **State Freight Mobility Funds** have committed an additional $4.6 million;

- **The Federal Highway Administration** provided $850,000 in demonstration funds;

- **The Port of Seattle** provided $5.5 million; and

- **Other funding sources** make up the remaining $500,000.

The Central Puget Sound region was designated as a maintenance area for carbon monoxide in 1996. The quantified air quality benefits of the SR 519 improvements will help the region meet its emissions budgets, an important element in ensuring that the project moves toward implementation.

5.2.3 COLTON CROSSING

The Colton Crossing is the intersection of the Union Pacific and Burlington Northern Santa Fe railroad tracks in Colton, California just south of I-10. Explosive growth in maritime freight traffic at the San Pedro Bay ports in Los Angeles has contributed to increasing rail congestion at this strategic intersection, because the majority of containers that move through the ports via rail must pass through the Colton Crossing. More than 110 trains pass through the crossing each day, making it one of the busiest rail intersections in the country. These tracks also are used for commuter trains. The growth in freight traffic has caused increased wait times for trains that must idle while waiting for other trains to pass. It also has led to passenger vehicle congestion at highway/rail grade crossings, which increases emissions.

> **KEY THEMES**
> - Rail/rail grade separation
> - Effective use of public-private partnerships

The proposed Colton Crossing grade separation would provide an east-west structure to separate the BNSF and UP tracks and allow for greater passenger and freight mobility. A benefit/cost analysis prepared for the railroads found that reduced locomotive and vehicle idling from implementing the grade separation would save nearly one million gallons of gasoline and diesel fuel annually. It also would reduce carbon dioxide emissions by about 34,000 tons per year, with significant reductions in other pollutants such as carbon monoxide, nitrous oxide, and particulate matter.[96] A similar analysis conducted by Cambridge Systematics, Inc. for the Riverside County Transportation Commission estimated public benefits on four key metrics, two of which relate to air quality. The study determined that a grade separation would reduce emissions from idling locomotives as well as passenger cars stopped at grade crossings.

In 2010, this project was awarded a USDOT TIGER Discretionary Grant for $33.8 million dollars. The remaining financing will come from a combination of California Trade Corridor Improvement Fund (TCIF) money and contributions from the UP and BNSF railroads. The total project cost is estimated at nearly $200 million dollars. The funding breakdown is as follows:

- $97 million from TCIF;

- $67 million from the BNSF and UP railroads; and

- $33.8 million from a USDOT TIGER Discretionary Grant.

---

[96] HDR/HLB Decision Economics, Inc., BNSF and Union Pacific Public Benefit Study for Colton Crossing Grade Separation, February 7, 2008.

Glossary of Terms

Air Pollutant – Any substance in air that could, in high enough concentration, harm man, other animals, vegetation, or material. Pollutants may include almost any natural or artificial composition of airborne matter capable of being airborne. They may be in the form of solid particles, liquid droplets, gases, or in combination thereof. Generally, they fall into two main groups (1) those emitted directly from identifiable sources and (2) those produced in the air by interaction between two or more primary pollutants, or by reaction with normal atmospheric constituents, with or without photoactivation. Air pollutants are often grouped in categories for ease in classification; some of the categories are solids, sulfur compounds, volatile organic chemicals, particulate matter, nitrogen compounds, oxygen compounds, halogen compounds, radioactive compound, and odors.

Air Pollution – The presence of contaminants or pollutant substances in the air that interfere with human health or welfare, or produce other harmful environmental effects.

Air Toxics – Any air pollutant for which a national ambient air quality standard (NAAQS) does not exist (i.e. excluding ozone, carbon monoxide, PM_{10}, sulfur dioxide, nitrogen oxide) that may reasonably be anticipated to cause cancer; respiratory, cardiovascular, or developmental effects; reproductive dysfunctions, neurological disorders, heritable gene mutations, or other serious or irreversible chronic or acute health effects in humans.

Airborne Particulates – Total suspended particulate matter found in the atmosphere as solid particles or liquid droplets. Chemical composition of particulates varies widely, depending on location and time of year. Sources of airborne particulates include dust, emissions from industrial processes, combustion products from the burning of wood and coal, combustion products associated with motor vehicle or non-road engine exhausts, and reactions to gases in the atmosphere.

Arterial – Major streets or highways, many with multilane or freeway design, serving high-volume traffic corridor movements that connect major generators of travel. While they may provide access to abutting land, their primary function is to serve traffic moving through the area.

Attainment Area – A geographic area in which levels of a criteria air pollutant meet the health-based primary standard (national ambient air quality standard, or NAAQS) for the pollutant. An area may have on acceptable level for one criteria air pollutant, but may have unacceptable levels for others. Thus, an area could be both attainment and nonattainment at the same time. Attainment areas are defined using federal pollutant limits set by regulatory agencies.

Backhaul – The process of a transportation vehicle (typically a truck) returning from the original destination point to the point of origin. A backhaul can be with a full or partially loaded trailer.

Barge – The cargo-carrying vehicle that inland water carriers primarily use. Basic barges have open tops, but there are covered barges for both dry and liquid cargoes.

Belly Cargo – Air freight carried in the belly of passenger aircraft.

Bottleneck – A section of a highway or rail network that experiences operational problems such as congestion. Bottlenecks may result from factors such as reduced roadway width or steep freeway grades that can slow trucks.

Breakbulk Cargo – Cargo of non-uniform sizes, often transported on pallets, sacks, drums, or bags. These cargoes require labor-intensive loading and unloading processes. Examples of breakbulk cargo include coffee beans, logs, or pulp.

Bulk Cargo – Cargo that is unbound as loaded; it is without count in a loose unpackaged form. Examples of bulk cargo include coal, grain, and petroleum products.

Cabotage – A national law that requires costal and intercostal traffic to be carried in its own nationally registered, and sometimes built and crewed ships.

Capacity – The physical facilities, personnel and process available to meet the product of service needs of the customers. Capacity generally refers to the maximum output or producing ability of a machine, a person, a process, a factory, a product, or a service.

Carbon Dioxide (CO_2) – A naturally occurring gas fixed by photosynthesis into organic matter. A by-product of fossil fuel combustion and biomass burning, it is also emitted from land-use changes and other industrial processes.

Carbon Monoxide (CO) – A colorless, odorless, poisonous gas, produced by incomplete burning of carbon-based fuels, including gasoline, oil, and wood. When carbon monoxide gets into the body, the carbon monoxide combines with chemicals in the blood and prevents the blood from bringing oxygen to cells, tissues, and organs. High-level exposures to carbon monoxide can cause serious health effects, with death possible from massive exposures. Symptoms of exposure to carbon monoxide can include vision problems, reduced alertness, and general reduction in mental and physical functions. Carbon monoxide exposures are especially harmful to people with heart, lung, and circulatory system diseases.

Carload – Quantity of freight (in tons) required to fill a railcar; amount normally required to qualify for a carload rate.

Carrier – A firm which transports goods or people via land, sea, or air.

Catalytic Converter – An air pollution abatement device that removes pollutants from motor vehicle exhaust, either by oxidizing them into carbon dioxide and water or reducing them to nitrogen.

Chassis – A trailer-type device with wheels constructed to accommodate containers, which are lifted on and off.

Class I Railroad – Railroads which have annual gross operating revenues over $266.7 million.

Class II Railroad – See Regional Railroad.

Class III Railroad – See Shortline Railroad.

Clean Air Act – Originally passed in 1963, although the 1970 version of the law is the basis of today's U.S. national air pollution program. The 1990 Clean Air Act Amendments are the most far-reaching revisions of the 1970 law, and are usually referred to as the 1990 Clean Air Act.

Climate Change (also referred to as "global climate change") – A change in the mean state or variability of the climate, whether due to natural variability or as a result of human activity, that persists for an extended period, typically decades or more. In some cases, "climate change" has been used synonymously with the term "global warming"; scientists however, tend to use the term in the wider sense to also include natural changes in climate.

Coastal Shipping – Also known as short-sea or coastwise shipping, describes marine shipping operations between ports along a single coast or involving a short sea crossing.

Combustion – Burning of fuels such as coal, oil, gas, and wood. Many important pollutants, such as sulfur dioxide, nitrogen oxides, and particulates (PM_{10}) are combustion products.

Commodity – An item that is traded in commerce. The term usually implies an undifferentiated product competing primarily on price and availability.

Common Carrier – Any carrier engaged in the interstate transportation of persons/property on a regular schedule at published rates, whose services are for hire to the general public.

Concentration – The relative amount of a substance mixed with another substance. Examples are 5 parts per million (ppm) of carbon monoxide in air.

Conformity – A process in which transportation plans and spending programs are reviewed to ensure they are consistent with federal clean air requirements; transportation projects collectively must not worsen air quality.

Container – A "box" typically ten to forty feet long, which is used primarily for ocean freight shipment. For travel to and from ports, containers are loaded onto truck chassis' or on railroad flatcars.

Container on Flatcar (COFC) – Containers resting on railway flatcars without a chassis underneath.

Containerization – A shipment method in which commodities are placed in containers, and after initial loading, the commodities per se are not rehandled in shipment until they are unloaded at destination.

Containerized Cargo – Cargo that is transported in containers that can be transferred easily from one transportation mode to another.

Criteria Air Pollutants – The 1970 amendments to the Clean Air Act required EPA to set National Ambient Air Quality Standards for certain pollutants known to be hazardous to human health. EPA has identified and set standards to protect human health and welfare for six pollutants ozone, carbon monoxide, total suspended particulates, sulfur dioxide, lead, and nitrogen oxide. The term, "criteria pollutants" derives from the requirement that EPA must describe the characteristics and potential health and welfare effects of these pollutants. It is on the basis of these criteria that standards are set or revised.

Deadhead – The return of an empty transportation container back to a transportation facility. Commonly-used description of an empty backhaul.

Distribution Center (DC) – The warehouse facility which holds inventory from manufacturing pending distribution to the appropriate stores.

Dock – A space used or receiving merchandise at a freight terminal.

Double-Stack – Railcar movement of containers stacked two high.

Drayage – Transporting of rail or ocean freight by truck to an intermediate or final destination; typically a charge for pickup/delivery of goods moving short distances (e.g., from marine terminal to warehouse).

Emission – Release of pollutants into the air from a source.

Emission Factor – The relationship between the amount of pollution produced and the amount of fuel consumed.

Emission Inventory – A listing, by source, of the amount of air pollutants discharged into the atmosphere of a community; used to establish emission standards.

Exhaust Gas Recirculation System (EGR) – The controlled diversion of some of the combustion gases back into the combustion chamber, lowering the combustion temperature and reducing nitrogen oxides in the engine. This is a very effective process, because oxides of nitrogen tend to rise disproportionately with increased combustion temperatures. There are two methods of exhaust gas recirculation: internally through overlap of valve opening times and externally with recirculation valves and manifolds.

Fossil Fuel – Fuel derived from ancient organic remains; e.g. peat, coal, crude oil, and natural gas.

Fuel Cell – An electrochemical engine (no moving parts) that converts the chemical energy of a fuel, such as hydrogen, and an oxidant, such as oxygen, directly to electricity. The principal components of a fuel cell are catalytically activated electrodes for the fuel (anode) and the oxidant (cathode) and an electrolyte to conduct ions between the two electrodes.

Greenhouse Effect – The process by which the absorption of infrared radiation by the atmosphere warms the Earth. In common parlance, the term "greenhouse effect" may be used to refer either to the natural greenhouse effect, due to naturally occurring greenhouse gases, or to the enhanced (anthropogenic, or man-made) greenhouse effect, which results from gases emitted by human activities.

Greenhouse Gas – A gas, whether natural or man-made, that contributes to the greenhouse effect by absorbing and emitting radiation at specific wavelengths within the spectrum of infrared radiation emitted by the Earth's surface, atmosphere, and clouds. Greenhouse gases include water vapor (H_2O), carbon dioxide (CO_2), nitrous oxide (N_2O), methane (CH_4), ozone (O_3), hydrofluorocarbons (HFC), and others.

Gross Vehicle Weight (GVW) – The combined total weight of a vehicle and its freight.

Hazardous Air Pollutants (HAPs) – Chemicals that cause serious health and environmental effects. Health effects include cancer, birth defects, and nervous system problems. HAPs are released by sources such as chemical plants, dry cleaners, printing plants, and motor vehicles (cars, trucks, buses, etc.).

Hydrocarbons (HC) – Chemical compounds that consist entirely of carbon and hydrogen.

Intermodal Terminal – A location where links between different transportation modes and networks connect. Using more than one mode of transportation in moving persons and goods. For example, a shipment moved over 1,000 miles could travel by truck for one portion of the trip, and then transfer to rail at a designated terminal.

Just-in-Time (JIT) – Cargo or components that must be at a destination at the exact time needed. The container or vehicle is the movable warehouse.

Laker – Large commercial ship operating on the Great Lakes carrying bulk cargo The Lakers are up to 1,000 feet long and can carry up to 66,000 tons of cargo. The large bulk Lakers stay within the Great Lakes because they are too large to enter the Saint Lawrence Seaway portion.

Lead (Pb) – A heavy metal that is hazardous to health if breathed or swallowed. Its use in gasoline, paints, and plumbing compounds has been sharply restricted or eliminated by federal laws and regulations.

Less-Than-Containerload/Less-Than-Truckload (LCL/LTL) – A container or trailer loaded with cargo from more than one shipper; loads that do not by themselves meet the container load or truckload requirements.

Level of Service (LOS) – A qualitative assessment of a road's operating conditions. For local government comprehensive planning purposes, level of service means an indicator of the extent or degree of service provided by, or proposed to be provided by, a facility based on and related to the operational characteristics of the facility. Level of service indicates the capacity per unit of demand for each public facility.

Lift-on/Lift-off (lo/lo) Cargo – Containerized cargo that must be lifted on and off vessels and other vehicles using handling equipment.

Line Haul – The movement of freight over the road/rail from origin terminal to destination terminal, usually over long distances.

Liquid Bulk Cargo – A type of bulk cargo that consists of liquid items, such as petroleum, water, or liquid natural gas.

Logistics – All activities involved in the management of product movement; delivering the right product from the right origin to the right destination, with the right quality and quantity, at the right schedule and price.

Methane – A colorless, nonpoisonous, flammable gas created by anaerobic decomposition of organic compounds. It is the main component of natural gas and is a greenhouse gas.

Mobile Sources – Moving objects that release pollution; mobile sources include cars, trucks, buses, planes, trains, motorcycles, and gasoline-powered lawn mowers. Mobile sources are divided into two groups: on-road vehicles, which include cars, trucks and buses, and nonroad vehicles, which includes trains, planes, lawn mowers, and some portable equipment.

Neo-bulk Cargo – Shipments consisting entirely of units of a single commodity, such as cars, lumber, or scrap metal.

Nitric Oxide (NO) – A gas formed by combustion under high temperature and high pressure in an internal combustion engine. NO is converted by sunlight and photochemical processes in ambient air to nitrogen oxide. NO is a precursor of ground-level ozone pollution, or smog.

Nitrogen Dioxide (NO_2) – The result of nitric oxide combining with oxygen in the atmosphere; major component of photochemical smog.

Nitrogen Oxides (NO_x) – A criteria air pollutant. Nitrogen oxides are produced from burning fuels, including gasoline and coal. Nitrogen oxides are smog formers, which react with volatile organic compounds to form smog. Nitrogen oxides are also major components of acid rain.

Nonattainment Area – A geographic area in which the level of a criteria air pollutant is higher than the level allowed by the federal standards. A single geographic area may have acceptable levels of one criteria air pollutant but unacceptable levels of one or more other criteria air pollutants; thus, an area can be both attainment and nonattainment at the same time.

On-dock Rail – Direct shipside rail service. Includes the ability to load and unload containers/breakbulk directly from rail car to vessel.

Owner-operator – Trucking operation in which the owner of the truck is also the driver.

Ozone – A gas composed of three oxygen atoms bound together into an ozone molecule (O_3). Ozone occurs in nature; it produces the sharp smell you notice near a lightning strike. High concentrations of ozone gas are found in a layer of the atmosphere – the stratosphere – high above the Earth. Stratospheric ozone shields the Earth against harmful rays from the sun, particularly ultraviolet B. Smog's main component is ozone; this ground-level ozone is a product of reactions among chemicals produced by burning coal, gasoline and other fuels, and chemicals found in products including solvents, paints, hairsprays, etc.

Particulates; Particulate Matter – A criteria air pollutant. Particulate matter includes dust, soot and other tiny bits of solid materials that are released into and move around in the air. Particulates are produced by many sources, including burning of diesel fuels by trucks and buses, incineration of garbage, mixing and application of fertilizers and pesticides, road construction, industrial processes such as steel making, mining operations, agricultural burning (field and slash burning), and operation of fireplaces and woodstoves. Particulate pollution can cause eye, nose and throat irritation, heart and lung disease, increased respiratory symptoms and disease, decreased lung function, and premature death.

Parts Per Billion (ppb)/Parts Per Million (ppm) – Units commonly used to express contamination ratios, as in establishing the maximum permissible amount of a contaminant in water, land, or air.

Payload – The cargo carried in a vehicle exclusive of the vehicle itself.

Piggyback – A rail/truck service. A shipper loads a highway trailer, and a carrier drives it to a rail terminal and loads it on a flatcar; the railroad moves the trailer-on-flatcar combination to the destination terminal, where the carrier offloads the trailer and delivers it to the consignee.

$PM_{10}/PM_{2.5}$ – PM_{10} is measure of particles in the atmosphere with a diameter of less than or equal to 10 micrometers. PM2.5 is a measure of particles less than or equal to 2.5 micrometers in diameter. Therefore, PM_{10} includes $PM_{2.5}$.

Pool/Drop Trailers – Trailer that are staged at a facilities for preloading purposes.

Port Authority – State or local government that owns, operates, or otherwise provides wharf, dock, and other terminal investments at ports.

Radio Frequency (RFID) – A form of wireless communication that lets users relay information via electronic energy waves from a terminal to a base station, which is linked in turn to a host computer. The terminals can be placed at a fixed station, mounted on a forklift truck, or carried in the worker's hand. The base station contains a transmitter and receiver for communication with the terminals. When combined with a bar-code system for identifying inventory items, a radio-frequency system can relay data instantly, thus updating inventory records in so-called "real time."

Rail Siding – A very short branch off a main railway line with only one point leading onto it. Sidings are used to allow faster trains to pass slower ones or to conduct maintenance.

Reefer Trailer – A refrigerated trailer that is commonly used for perishable goods.

Regional Railroad – Railroad defined as line-haul railroad operating at least 350 miles of track and/or earns revenue between $40 million and $266.7 million.

Reliability – Refers to the degree of certainty and predictability in travel times on the transportation system. Reliable transportation systems offer some assurance of attaining a given destination within a reasonable range of an expected time. An unreliable transportation system is subject to unexpected delays, increasing costs for system users.

Roll-on/Roll-off (ro/ro) Cargo – Wheeled cargo, such as automobiles, or cargo carried on chassis that can be rolled on or off vehicles without using cargo handling equipment.

Short Line Railroad – Freight railroads which are not Class I or Regional Railroads, that operate less than 350 miles of track and earn less than $40 million.

Short-sea Shipping – Also known as coastal or coastwise shipping, describes marine shipping operations between ports along a single coast or involving a short sea crossing.

Smog – A mixture of pollutants, principally ground-level ozone, produced by chemical reactions in the air involving smog-forming chemicals. A major portion of smog-formers comes from burning of petroleum-based fuels such as gasoline. Other smog-formers, volatile organic compounds, are found in products such as paints and solvents. Smog can harm health, damage the environment and cause poor visibility. Major smog occurrences are often linked to heavy motor vehicle traffic, sunshine, high temperatures and calm winds or temperature inversion (weather condition in which warm air is trapped close to the ground instead of rising). Smog is often worse away from the source of the smog-forming chemicals, since the chemical reactions that result in smog occur in the sky while the reacting chemicals are being blown away from their sources by winds.

State Implementation Plan (SIP) – A detailed description of the programs a state will use to carry out its responsibilities under the Clean Air Act. State implementation plans are collections of the regulations used by a state to reduce air pollution in nonattainment areas. The Clean Air Act requires that EPA approve each state implementation plan. Members of the public are given opportunities to participate in review and approval of state implementation plans.

Sulfur Dioxide – A criteria air pollutant and gas produced by burning sulfur-containing fuels. Coal combustion (particularly for power plants) is the largest source, but it is also produced by paper production, metal smelting, and diesel fuel combustion. Sulfur dioxide is closely related to sulfuric acid, a strong acid. Sulfur dioxide plays an important role in the production of acid rain.

Switching and Terminal Railroad – Railroad that provides pick-up and delivery services to line-haul carriers.

Temperature Inversion – One of the weather conditions that are often associated with serious smog episodes in some portions of the country. In a temperature inversion, air does not rise because it is trapped near the ground by a layer of warmer air above it. Pollutants, especially smog and smog-forming chemicals, including volatile organic compounds, are trapped close to the ground.

Throughput – Total amount of freight imported or exported through a seaport measured in tons or TEUs.

Ton-mile – A measure of output for freight transportation; reflects weight of shipment and the distance it is hauled; a multiplication of tons hauled by the distance traveled. One ton of cargo transported one mile equals one ton-mile.

Trailer on Flatcar (TOFC) – Transport of trailers with their loads on specially designed rail cars.

Transloading – Transferring bulk shipments from the vehicle/container of one mode to that of another at a terminal interchange point.

Transshipment – Transferring any shipments from the vehicle/container of one mode to that of another at a terminal interchange point.

Truckload (TL) – Quantity of freight required to fill a truck, or at a minimum, the amount required to qualify for a truckload rate.

Twenty-foot Equivalent Unit (TEU) – The standard measure used for containerized cargo. The amount of cargo that would fill an eight-foot by eight-foot by 20-foot intermodal container.

Unit Train – A train of a specified number of railcars handling a single commodity type which remain as a unit for a designated destination or until a change in routing is made.

Vehicle Miles of Travel (VMT) – Each mile traveled by vehicle without regard to the contents of the vehicle. For example, a five-mile truck trip would generate five vehicle miles of travel.

Volatile Organic Compounds (VOCs) – Chemicals that produce vapors readily. At room temperature and normal atmospheric pressure, vapors escape easily from volatile liquid chemicals. Volatile organic chemicals include gasoline, industrial chemicals such as benzene, solvents such as toluene and xylene, and tetrachloroethylene (perchloroethylene, the principal dry cleaning solvent). Many volatile organic chemicals are also hazardous air pollutants; for example, benzene causes cancer.

Weigh-in-Motion – Defined by the American Society for Testing and Materials (ASTM) as "the process of measuring the dynamic tire forces of a moving vehicle and estimating the corresponding tire loads of the static vehicle." It allows truck weights to be determined without requiring the vehicle to stop.